INDUSTRY AND AIR POWER

INDUSTRY
AND
AIR POWER
The Expansion of British
Aircraft Production,
1935-41

SEBASTIAN RITCHIE

Routledge
Taylor & Francis Group

LONDON AND NEW YORK

First published in 1997 in Great Britain by
Routledge
2 Park Square, Milton Park, Abingdon, Oxon, OX14 4RN
605 Third Avenue, New York, NY 10017

Routledge is an imprint of the Taylor & Francis Group,
an informa business

British Library Cataloguing in Publication Data

Ritchie, Sebastian
    Industry and air power : the expansion of British aircraft
    production, 1935–41. – (Studies in air power)
    1. Aircraft industry – Great Britain – History
    I. Title
    338.4'7'6291'0941

Library of Congress Cataloging-in-Publication Data

Ritchie, Sebastian, 1963–
    Industry and air power : the expansion of British aircraft
    production, 1935–41 / Sebastian Ritchie.
        p.   cm. — (Routledge series—studies in air power)
    Includes bibliographical references and index.
    ISBN 0-7146-4724-1 (cloth) – ISBN 0-7146-4343-2 (pbk.)
    1. Aircraft industry—Military aspects—Great Britain—History.
    2. Airplanes, Military—Great Britain—History.   I. Title.
    II. Series.
    HD9711.G72R57   1996
    338.4'762913'094109043—dc20   96-43588
                                                                CIP

Typeset by Regent Typesetting, London

Publisher's Note
The publisher has gone to great lengths to ensure the quality of this reprint but points out that some imperfections in the original may be apparent

ISBN 13: 978-0-7146-4343-4 (pbk)
ISBN 13: 978-0-7146-4724-1 (hbk)

For Owen, and in Memory of Paddy

Contents

List of tables and figures

TABLES

FIGURES

Editor's Foreword

The editorial intention of the Cass Air Power Studies Series is to explore all aspects of air power, including themes and topics which have previously been neglected. The economic and industrial foundation of air power is one such theme, and Dr Ritchie's work is among the first to study the subject in depth.

The period immediately before the Second World War was a particularly crucial one, which saw the industrial capacity put in place to underpin Britain's air effort during the conflict. The author examines the relationship between industry and the state at a time when increasing tension resulted in large government contracts. Simultaneously, however, conflicting political aims, for example the immediate desire for large 'shop-window' deterrent forces as opposed to the longer-term need for war-capable air forces, produced contradictions in official policy. These in turn induced strains in the relationship between government and industry, and within the machinery of state itself. Dr Ritchie explores these varying factors using not only the official papers on which much existing scholarship has been based, but also surviving company papers. The result is an admirably balanced account.

He also challenges some more recent scholarship which has sought to suggest that the aircraft industry immediately before the war was a paradigm for British inefficiency and industrial decline. He argues instead that there is much to be admired in the policies and achievements described here.

SEBASTIAN COX

Acknowledgements

In preparing this book, I was given access to the archives of three private companies: the Rolls-Royce archive at Derby, the Vickers archive at Cambridge University Library, and the Hawker Siddeley archive at Slough. I was also allowed to consult the archives of the Engineering Employers Federation and those of the British Motor Industry Heritage Trust. I wish to acknowledge with thanks the assistance given to me by all of these organisations, and their permission to reproduce material from their archives.

I would also like to thank the series editor, Sebastian Cox, and Professor Richard Overy of King's College London, for the invaluable help and advice they provided during the preparation of this book.

Sebastian Ritchie

Abbreviations

Abbreviations for archival sources are as follows:

AC – Royal Air Force Museum Archive (Handley Page and Bristol Records)

EEFAW – Engineering Employers Federation Archive, Modern Records Centre, University of Warwick

EEFAL – Engineering Employers Federation Archive, London

HP – Handley Page Archive, Royal Air Force Museum

HSA – Hawker Siddeley Archive

PRO – Public Record Office

ROBN – Professor Sir Austin Robinson papers

RRA – Rolls-Royce Archive

USSBS – United States Strategic Bombing Survey

Vickers – Vickers Archive

Weir – Lord Weir papers

Abbreviations for committee minutes cited in the notes are as follows:

ACCS – Air Council Subcommittee on Supply

ACM NTC – Aircraft Manufacturers National Technical Committee of Engineering Employers Federation

AESC – Aero-Engine Shadow Committee

ASB – Air Supply Board

DGP – Director General of Production's meetings

DRC – Ministerial Committee on Defence Requirements

S of S EPM – Secretary of State for Air's RAF Expansion Progress Meeting

Abbreviations for Air Ministry and Ministry of Aircraft Production Personnel cited in the notes are as follows:

AD – Assistant Director

AMDP – Air Member for Development and Production

AMRD – Air Member for Research and Development

AMSO – Air Member for Supply and Organisation
AMSR – Air Member for Supply and Research
CAS – Chief of the Air Staff
DAP – Director of Aeronautical Production
DCF – Director of Capital Finance
DD – Deputy Director
DGAP – Director General of Aircraft Production
DGE – Director General of Equipment
DGP – Director General of Production
DGPS – Director General of Programmes and Statistics
D of C – Director of Contracts
D of E – Director of Equipment
DS – Deputy Secretary
D Stats. P – Director of Statistics and Programmes
DUS – Deputy Under-Secretary
DTD – Director of Technical Development
PAS – Principal Assistant Secretary
PS – Private Secretary
PSO - Principal Supply Officer
PUS – Permanent Under-Secretary
S of S – Secretary of State
US of S – Under-Secretary of State

Other abbreviations:
EEF – Engineering Employers Federation
MAP – Ministry of Aircraft Production
SBAC – Society of British Aircraft Constructors
AEU – Amalgamated Engineering Union
MAF – Metropolitan Air Force
ACGB – Aeronautical Corporation of Great Britain
APD – Armaments Profits Duty
EPT – Excess Profits Tax

Introduction

This is both a military and an industrial history. It is a military history in that it examines state armaments procurement policy but it is also intended to contribute to the history of the armaments manufacturing industry. Its subject is one of the largest state-sponsored industrial enterprises in British history: the expansion of the military aircraft industry in the rearmament years of the late 1930s and in the early years of the Second World War.

This is not the first history of the expansion of the aircraft industry but it is the first to be devoted entirely to that subject. The official civil histories of the Second World War were more concerned with state policy than with the performance of industry. Moreover, they dealt with each different aspect of state policy in relation to the armaments industry as a whole.[1] Other works have examined the relationship between the airframe industry and the state, and the military aspects of British air policy between the wars.[2] Studies have also been made of the political and financial consequences of rearmament.[3] However, no individual survey of aircraft production in this period has ever been prepared.

This account therefore fills a gap in the history of both air rearmament and the aircraft industry by examining state production policy, industrial production, employment and finance in the context of that industry alone. It is also based on a far wider range of sources than most earlier works which have tended to rely overwhelmingly on the state archives. The official records may well be an invaluable historical source for students of public administration but they should not be expected to provide a reliable insight into the affairs of government contractors such as aircraft companies. Industrial problems must ultimately be examined at the industrial level if they are to be properly understood and for this reason it is impossible to do justice to the history of rearmament without carefully scrutinising the aircraft industry's own archives and those of its trade and employment organisations.[4]

For many years a clear contradiction existed between the popular

perceptions of the success of the wartime aircraft economy and the
reality of Britain's protracted decline as an industrial country.
Success in a technologically advanced manufacturing sector
appeared entirely inconsistent with the general pattern of industrial
failure which has characterised Britain's economic performance
since 1945. More recently, Correlli Barnett has attempted to re-
interpret the history of wartime aircraft production in the light of
this failure. Given the fact of Britain's post-war economic decline, it
was argued, the aircraft industry could not have been as successful
as historians had previously thought. The deficiencies of wartime
aircraft production were ultimately of much greater historical
significance than the achievements, for they were caused by the
same structural weaknesses that have progressively undermined
Britain's industrial vitality in the twentieth century.[5]

However, the twin questions of continuity and discontinuity are
far more complex than this argument suggests, for two reasons.
First, the circumstances in which industry has to operate in wartime
are often very different from those obtaining in peacetime. War may
impose many of its own constraints on industrial production but,
conversely, wartime conditions may also offer solutions to problems
which might not so easily be solved in peacetime. There is, for
example, more scope for co-operation between different companies
and between management and labour than might be expected in
peacetime. Moreover, in wartime the role of the state assumes a
critical importance. The effectiveness of the official procurement
apparatus can make an enormous difference to the performance of
the industrial enterprise as a whole.

Second, the declinists' view of unmitigated post-war industrial
failure is consistently overstated. Had the performance of British
industry been as universally discreditable as they believe the
economy would almost certainly have collapsed altogether. It has
not done so: it has declined relatively but not absolutely. Indeed,
post-war rates of output and productivity growth compare
favourably with those recorded in the late-nineteenth and early
twentieth centuries. Britain produces many more trained scientists
and engineers than at any previous time in her industrial history,
and her industries invest much more heavily in fixed capital and in
research and development.

It may be, however, that some sectors of British industry have
performed better than others, and there are good reasons why the
military aircraft industry should have been among the more

successful. Throughout its history it has received more generous state support than any other sector of British manufacturing industry; it has always employed a disproportionately high number of trained scientists, technicians and engineers; and it has enjoyed unusually close relations with foreign companies from the earliest years of its existence, when technology flowed freely between the aeronautical powers, until the present with its multi-national collaborative aerospace ventures.[6] In short, the fact that Britain has declined relatively as an industrial country in the twentieth century does not in itself prove that the wartime aircraft economy was very inefficient. It is all too easy in the context of Britain's post-war economic problems to explain why Britain did not produce more aircraft during the war. It is much more difficult to show how British industry managed to produce as many aircraft as it did. The objective here is to resolve precisely that paradox.

Air rearmament did not begin in 1935, and the expansion of the aircraft industry did not come to an end in 1941, but the 1935–41 period nevertheless suggested itself for several reasons. In 1935 the first major rearmament plan, scheme C, was launched by the National Government. Scheme C was the first plan to affect the entire aircraft industry, to necessitate a general expansion of industrial capacity, and to order, albeit in small quantities, aircraft incorporating the radical new technology of the all-metal monoplane. The year 1941 was the target date for the single most important aircraft rearmament plan – the War Potential programme. This programme was initiated in 1938 with the aim of giving Britain the capacity to produce 2000 aircraft per month by the end of 1941. Until the programme was sanctioned British aircraft production essentially entailed rearmament 'in breadth' by providing for the creation of a particular Royal Air Force (RAF) first-line strength by a particular date. But the War Potential programme provided a basis for planning aircraft production in much greater depth and for developing a comprehensive state production organisation. Just as the programme was nearing its completion a new bomber production plan was launched. From then on, as Sir Alec Cairncross has recently written, 'the shape of the aircraft programme was set for the rest of the war'.[7]

The rearmament years, however, cannot be artificially segregated from the inter-war period as a whole. The first chapter therefore surveys the history of the aircraft industry and of state procurement policy between 1918 and 1935, and focuses on the industry's structure, its profitability, the volume of its operations and its technical competence. The Air Ministry's thinking on wartime industrial mobilisation is also analysed.

Chapter 2 examines the Air Ministry's production organisation and policy between 1935 and the outbreak of war, and identifies two main phases of policy making. In the first, from May 1935 to March 1938, the aims of air rearmament were strictly limited and state production policy and organisation were consequently slow to develop. During the second, from March 1938 to September 1939, the expansion plans were extended to prepare for the event of war. As the scale and complexity of the aircraft programme increased, the Air Ministry was forced to accept formidable new administrative burdens.

The airframe industry was the largest single sector of the aircraft industry, with responsibility for both designing and assembling complete aircraft, and for manufacturing airframes. Chapter 3 considers the airframe industry's response to air rearmament. It argues that technical standardisation was an essential for quantity production, and that delays in achieving the necessary degree of standardisation were primarily the result of the technological revolution in aircraft design – the replacement of the biplane by the monoplane – which occurred in the mid-1930s. This chapter also assesses the scale of investment in new factories and plant and examines some of the new design and production methods introduced by the industry to cope with the growing volume of production.

Chapter 4 deals with the single most important sector of the industry: the aero-engine sector. Here the problem of technical change, although challenging, was less acute than in the airframe sector, but the pressure of large-scale production did force aero-engine firms like Rolls-Royce to reorganise themselves and led the government to seek the assistance of the motor car industry to achieve the necessary increase in output. This account draws extensively on industrial sources to consider the relationship between rearmament and corporate modernisation and to describe how the productive base of the professional aero-engine companies was enlarged by transplanting technology and specialised manufacturing processes from one industry to another.

Chapters 5 and 6 focus on two particular industrial resources: manpower and finance. The aim in Chapter 5 is to assess the adequacy of the manpower supply, and the extent to which supply problems could be surmounted by better use of manpower. Chapter 6 provides an account of the aircraft boom: the remarkable stampede for aircraft shares which raised much of the capital required by the industry during the early stages of expansion. It also examines the role of the state in financing aircraft production, contractual relations between the state and the aircraft industry, and the search for a contractual formula which not only controlled costs but also encouraged higher output.

The Air Ministry's war production plans were necessarily devised on a somewhat hypothetical basis so it was inevitable that pre-war assumptions about the pace of industrial mobilisation and the direction of technical development should have been reconsidered in the light of combat experience. Chapter 7 describes the administrative changes and the planning and investment decisions implemented during the first two years of hostilities in order to turn the aircraft programme into a form better suited to wartime requirements. It also surveys the progress of production, the constraints on output, and the efforts of the state and the aircraft industry to improve the efficiency with which resources were employed. Finally it assesses the effectiveness of the structure and organisation of the wartime aircraft industry.

One fundamental truth needs to be borne in mind: the emergency expansion of an advanced manufacturing industry involves scientific, managerial and engineering problems of extraordinary complexity. Few historians are professionally qualified to comprehend or explain these issues adequately, yet to write an industrial history without the necessary degree of scientific understanding is to risk interpretative error. My aim has been to seek precise, technical, explanations for the industrial problems encountered during the rearmament year, and not to ascribe them to some kind of deep-rooted national antipathy to industry and technology, or to such ill-defined concepts as managerial inefficiency or Air Ministry conservatism. It is for this reason more than any other that my conclusions differ from those reached in most previous histories of British military aircraft production.

NOTES

1. For the most detailed work on aircraft production in the official series see W. Hornby, *Factories and Plant* (London, 1958). See also M. Postan, *British War Production* (London, 1952); W. Ashworth, *Contracts and Finance* (London, 1953); P. Inman, *Labour in the*

Munitions Industries (London, 1957); M. Postan, D. Hay, and J. Scott, *Design and Development of Weapons* (London, 1964); J. Scott and R. Hughes, *The Administration of War Production* (London, 1955). For a more recent analysis of state policy relating specifically to wartime aircraft production see Sir Alec Cairncross, *Planning in Wartime: Aircraft Production in Britain, Germany and America* (London, 1991).

2. On the impact of state policy on inter-war aircraft manufacturing see P. Fearon, 'The British Airframe Industry and the State, 1918–35', *Economic History Review*, 27 (1974); for a critique of Fearon's thesis see A. Robertson, 'The British Airframe Industry and the State in the Interwar Period: A Comment', *Economic History Review*, 28 (1975). On air strategy see M. Smith, *British Air Strategy between the Wars* (Oxford, 1984).

3. R. Shay, *British Rearmament in the Thirties* (Princeton, 1977); G. Peden, *British Rearmament and the Treasury 1932–1939* (Edinburgh, 1979).

4. To date the two best studies based on the aircraft industry's own archives are: P. Fearon, 'The Vicissitudes of a British Aircraft Company: Handley Page Ltd. Between the Wars', *Business History*, 20 (1978); I. Lloyd, *Rolls-Royce: The Merlin at War* (London, 1978), and I. Lloyd, *Rolls-Royce: The Years of Endeavour* (London, 1978). Fearon assumed that the pattern of Handley Page's development mirrored that of the aircraft industry as a whole, but evidence drawn from the other firms and presented below (see Chapter 1) does not substantiate this view. Lloyd's account is a company history and does not compare Rolls-Royce's experiences with those of other aircraft or aero-engine firms.

5. C. Barnett, *The Audit of War: The Illusion and Reality of Britain as a Great Nation* (London, 1986), p. xi, Chapters 3, 7 and 8. For a critique of Barnett's thesis see D. Edgerton, 'The Prophet Militant and Industrial: The Peculiarities of Correlli Barnett', *Twentieth Century British History*, Vol. 2, No. 3 (1991). The efficiency and overall success of Britain's wartime aircraft economy is assessed throughout the following study, but Chapter 7 and Chapter 8 are particularly relevant to the debate on Barnett's work.

6. D. Edgerton, *England and the Aeroplane: An Essay on a Militant and Technological Nation* (London, 1991), *passim*; K. Hayward, *The British Aircraft Industry* (Manchester, 1989), pp. 131, 189–91, 203–6.

7. Cairncross, *Planning in Wartime*, p. 9.

1

British aircraft production
between the wars

The British aircraft industry was born in the decade preceding the outbreak of the First World War. Short Brothers and Handley Page were formed in 1908, Blackburn and A.V. Roe in 1910, Sopwith and British and Colonial (which later became Bristol) in 1911. Established armaments manufacturers like Vickers and Armstrong Whitworth also began designing and building aircraft in this period. Although private flying was initially the most important source of demand, the armed services' growing interest in the military potential of aviation led them to place substantial orders between 1912 and 1914, and at the start of the war the naval wing of the Royal Flying Corps (RFC) had a total of 93 heavier-than-air craft, while the military wing had 179. After this, service requirements resulted in a dramatic expansion of industrial capacity and output. A fivefold increase in employment in the aircraft industry occurred between October 1916 and 1918. Monthly production rose from about ten at the beginning of the war to 2500 in 1918.[1]

After the armistice, military demand inevitably contracted yet the newly formed RAF continued to provide a market for the vast majority of aircraft produced in Britain in the inter-war years. The equipment and technical development directorates of the Air Ministry's Department of Supply and Research and the contracts directorate of the Ministry's secretariat presided over RAF procurement. During the 1920s the Ministry adopted a policy of rationing design contracts between 14 different airframe firms, 11 of which were wholly or predominantly dependent on the Air Ministry for their survival. The intention was to maintain a nucleus of technical and productive capacity which could be expanded in wartime, to improve design through competition, and to prevent contractors from demanding excessive prices for their aircraft.

This chapter examines the consequences of the Air Ministry's policy. How did it affect the industry's structural characteristics, its profitability and the volume of its operations? How satisfied was the Air Ministry that a sufficient basis for wartime expansion existed and how did Air Ministry patronage influence the design and production techniques employed by the aircraft industry?

THE AIRCRAFT INDUSTRY AND AIRCRAFT PRODUCTION, 1918–35

The 'lean years'

In the history of the British aircraft industry it is usual for the period between the end of the First World War and the beginning of rearmament to be characterised as the 'lean years'.[2] During the war the industry was, of course, extremely profitable. Sopwith, the most famous firm of all, emerged in 1918 with reserves totalling more than £1 million. However, the post-war contraction of demand was accompanied by exceptional financial insecurity. Half of Sopwith's reserves were given up in excess profits duty and the firm was finally pushed into liquidation following a disastrous effort to diversify its activities.[3] Companies like Handley Page, which attempted to join the automobile industry after the war, were likewise hit by the recession of 1920–21. Efforts to promote foreign sales and to launch an air transport service proved equally fruitless, and the losses incurred by Handley Page from these activities totalled £606,000 in 1920.[4] Excess Profits Duty forced Martinsyde and the Aircraft Manufacturing Company out of business, and British and Colonial only managed to avoid the duty by transferring its assets to the Bristol Aeroplane Company.[5]

Many firms from outside the professional industry had been drawn into the aircraft economy during the First World War, but the depressed post-war market for aircraft persuaded the majority to return to their former activities. Some design firms also contemplated completely ceasing aircraft work. The board minutes of Sopwith Aviation record that in October 1919 'consideration was given to the advisability of the company totally abandoning aircraft'. However, 'it was agreed that although the market for aircraft during the next few years was likely to be small, if this company abandoned aircraft now it would be impossible to take it up again in the future. It was therefore resolved to continue designing and experimental work.' By the time Sopwith went into liquidation in 1920, Hawker Engineering Ltd had already been formed to continue with Sopwith design and manufacture.[6]

Following the government's decision, in 1923, to create a 52-squadron home defence force, the aircraft industry moved from what Professor Higham has described as 'demobilisational instability' to 'peacetime equilibrium', and the Air Ministry adopted its policy of rationing work between different contractors.[7] No accurate figures exist for total British aircraft production until the year 1930, when 1456 airframes were built, but Air Ministry orders, which included aircraft 'exported' to RAF overseas commands,[8] averaged 646 aircraft per year between 1923 and 1930 (see Table 1).

TABLE 1

British Aircraft Production, 1924–35

Year	Total Air Ministry Orders	RAF Home Commands Orders	Exports	Total UK Production
	A	B	C	D
1924	563	–	188	–
1925	448	–	148	–
1926	805	–	150	–
1927	392	–	140	–
1928	835	495	358	–
1929	615	573	525	–
1930	864	855	317	1,456
1931	–	728	304	–
1932	–	445	300	–
1933	–	633	234	–
1934	652	–	298	1,108
1935	893	–	453	1,807

Source: (A) PRO AIR 2/1322; PRO AIR 19/524, 1934–35 figures for deliveries to the Air Ministry; (B) M.M. Postan, *British War Production*, p. 5; (C) *Annual Statements of Trade of the UK*; (D) *Census of Production* (1930 and 1935).

The records show that most of the military airframe companies received at least £1.8 million from the Air Ministry in this period, and the accounts of one such company, Supermarine, demonstrate that it was possible to operate very profitably on the basis of such receipts (see Tables 2 and 3). Nevertheless, financial stability ultimately depended on success in consecutive design contests. Competition was strong because there were so many different contractors, and the chances of failure were correspondingly high. The problem was worsened by the tendency of some firms to rely on the success of their existing designs at the expense of research and development for the future. Hence Fairey Aviation, the most prosperous airframe firm of the 1920s, went public in 1929 with a capital of £500,000 only to see its profits collapse in the early 1930s (see Table 4).

TABLE 2

Air Ministry Payments (in £000) to Aircraft Firms

Airframe firms receiving less than £1.8 million, 1923–30								
Year	*1923*	*1924*	*1925*	*1926*	*1927*	*1928*	*1929*	*1930*
Boulton Paul	65	70	33	33	34	92	124	60
Gloster	90	309	232	314	197	166	108	80
Handley Page	71	64	152	167	188	172	155	205
Saunders	39	49	32	86	30	24	20	20

Airframe firms receiving more than £1.8 million, 1923–30								
Year	*1923*	*1924*	*1925*	*1926*	*1927*	*1928*	*1929*	*1930*
Short	39	73	105	109	56	51	53	109
Armstrong Whit.	77	92	121	195	330	413	419	467
Blackburn	208	321	299	250	145	194	226	281
Bristol Airframe	201	371	346	237	151	243	157	276
Hawker	190	217	336	408	216	86	123	372
A.V. Roe	240	281	345	279	261	131	99	172
Supermarine	129	97	167	308	308	260	332	248
Westland	115	114	142	166	64	227	513	536

Airframe firms with largest market share								
Year	*1923*	*1924*	*1925*	*1926*	*1927*	*1928*	*1929*	*1930*
Fairey	457	690	880	545	497	706	775	915
Vickers	283	689	493	493	420	416	522	395

Aero-engine firms								
Year	*1923*	*1924*	*1925*	*1926*	*1927*	*1928*	*1929*	*1930*
Armstrong Sid.	183	254	300	263	601	357	427	408
Bristol Engines	149	174	158	152	234	338	548	743
Napier	714	827	636	427	631	676	858	595
Rolls-Royce	188	422	433	231	403	299	248	459

Source: PRO AIR 2/1322.

TABLE 3

Supermarine Profits, Sales and Capital (£), 1923–27

Year	Sales	Capital Employed	Profit	% Profit to Sales	% Profit to Capital
1923	137,683	37,485	58,002	42.0	154.5
1924	192,525	52,642	68,371	35.5	130.0
1925	175,565	59,236	32,579	18.5	55.0
1926	267,243	61,680	77,816	29.0	126.0
1927	403,868	232,664	111,935	27.5	48.0

Source: Vickers K757.

TABLE 4

Net Profits (£) of Five Aircraft Firms

Year	Handley Page	Fairey	Westland	Armstrong Whitworth	Napier
1930	43,147	169,964	128,715	98,946	169,905
1931	3,594	184,585	118,530	89,673	80,233
1932	8,135	198,510	11,933	19,808	17,560
1933	20,441	115,477	3,994	54,078	−1,235
1934	44,590	47,034	−2,454	−21,023	21,093
1935	77,893	39,113	n.a.	−17,296	−46,831

Source: Annual Reports in *The Statist*; Handley Page archive; HSA, Armstrong Whitworth Accounts.

The financial status of the industry as a whole improved after 1923, but Air Ministry payments to five contractors were substantially lower than the average. One such firm was Handley Page, whose Air Ministry receipts grew steadily throughout the period. Another was Gloster, who was paid £314,000 in 1926 and just £80,000 in 1930. But the receipts of Short Brothers, Saunders, and Boulton Paul were lower still. These firms obtained little or no encouragement from the Air Ministry and only survived by diversifying their activities and using the income from non-aviation business to finance aircraft design and experimental work.[9] Between 1920 and 1930 Short built only 36 aircraft;[10] Saunders' average annual receipts from the Air Ministry amounted to just £37,250 between 1923 and 1930; and although Boulton Paul employed several thousand workers in 1925, only 150 were engaged in aircraft manufacture.[11] It is difficult to see such companies as serious contenders for the military market yet as long as they continued to undertake experimental work they could capitalise on any revival of air force demand whereas, as the Sopwith minutes suggest, abandoning any aircraft manufacturing meant abandoning aircraft manufacturing altogether.[12]

The financial problems facing these firms were exacerbated by the transition from wood to metal construction which occurred in the later 1920s. The new construction methods involved an increased commitment to development and testing, the acquisition of new machinery, and changes in manufacturing processes, each of which required considerable expenditure. Saunders was taken over in 1927 and made losses in the ensuing five years.[13] Even more affluent concerns like Blackburn were forced to seek external financial assistance 'as no shareholders had offered to find additional capital during a critical period'. In consequence the chairman, Robert Blackburn, lost his controlling rights on the board of directors, although he regained them in December 1931 after an upturn in the demand for the company's aircraft.[14]

The aggregate domestic demand for aircraft remained relatively stable until the beginning of rearmament but production contracts were less evenly rationed during the early 1930s than in the previous decade.[15] The export market also contracted. The number of aircraft exported fell from 525 in 1929 to 234 in 1933.[16] Moreover, firms became embroiled in another technical revolution: the replacement of the biplane by the monoplane. A memorandum from the industry's trade organisation, the Society of British Aircraft Constructors (SBAC), noted that 'since the transposition from wood and metal machines to all metal machines, and since the development of the art has required greater refinement and form, the cost of developing new machines has rapidly increased'.

A growing proportion of this expenditure was borne by the industry itself. The SBAC calculated that unrecovered expenditure on military research and development amounted to approximately five per cent of the industry's Air Ministry sales between 1930 and 1934.[17] Yet, in the absence of production orders, the all-important technical teams became, in the words of one industrialist, 'an unremunerative overhead charge'.[18] Frederick Handley Page was unable to stop his draughtsmen being poached by Vickers, who could afford to pay higher wages.[19]

At the same time the Air Ministry began to make detailed investigations of aircraft costs using the Admiralty's Technical Costs

Branch and, while these estimates were far from perfect, they undoubtedly helped to reduce profit margins still further.[20] Profits were also forced down by the Air Ministry's practice of awarding contracts to the cheapest bidders. Armstrong Whitworth, who produced Hawker's Hart bomber in the early 1930s, saw profits decline by 78 per cent between 1931 and 1932, and in the following three years the company made a loss. In November 1933 the chairman, Sir John Siddeley, sacked his general manager and launched an exhaustive investigation into Armstrong Whitworth's works organisation.[21] Several reforms were implemented but in March 1935 Siddeley had still to inform his directors that while 'the works were fully occupied . . . unfortunately the production was not of a profitable nature'.[22] Table 4 illustrates the financial problems encountered by some of the aircraft firms in the years prior to re-armament.

The trend towards consolidation

For some aircraft companies, then, the 1920s and early 1930s were indeed lean years, and historians like Peter Fearon have suggested that the experiences of these firms were representative of the industry as a whole. This is, however, a somewhat pessimistic view which does not stand up to detailed scrutiny, for an equally important feature of the industry's inter-war history was a trend towards consolidation which led a number of firms to expand their operations. Vickers purchased Supermarine in 1928; the Armstrong Siddeley Development Company (which included Armstrong Whitworth and Armstrong Siddeley Motors) acquired both A.V. Roe and the aluminium alloy producer, High Duty Alloys, in the same year, and was absorbed into the Hawker Siddeley group with Hawker and Gloster in 1935.

These firms increasingly dominated the military market. Of all the military aircraft built for the Air Ministry between January 1933 and December 1935, 78 per cent were either produced by Vickers, A.V. Roe, or Hawker, or else were designed by Hawker and subcontracted to other firms, including Vickers and Armstrong Siddeley subsidiaries.[23] By the early 1930s aero-engine manufacture was also concentrated in the hands of two firms, Bristol and Rolls-Royce, and the private flying market had been captured by De Havilland. An analysis of this process reveals that, contrary to the 'lean years' interpretation, aircraft manufacturing in inter-war Britain could be a lucrative business, and that there were some opportunities for the industry to gain experience in quantity production methods.

The Vickers-Supermarine merger

Vickers Aviation was Britain's largest inter-war aircraft manufacturer. Average annual receipts from the Air Ministry amounting to £464,000 between 1923 and 1930, were chiefly derived from the manufacture of large bomber aircraft, but it was widely held in the later 1920s that the future of large aircraft lay with the flying boat and, in 1928, Vickers decided to gain access to this market by purchasing one of the established flying boat firms.

Enquiries revealed that Blackburn's factories were 'not in a flourishing condition', Short's aviation work was too 'intermixed with other activities', and Saunders wanted an unreasonable sum for its business.[24]

In certain respects, however, Supermarine did not represent a much better proposition. Vickers' staff were critical of Supermarine's works organisation; the company's Southampton factory was inadequately equipped for the manufacture of the all-metal seaplanes which Vickers wanted to construct; wages at Supermarine were higher than those paid by Vickers, and the ratio of unskilled to skilled labour was 1:3 whereas at Vickers it was 3:1. Very little progress had been made on metal construction, and Vickers considered 'that the Supermarine company have a difficult and expensive period before them ... involving as it does experimental work, new designs, rearrangement of shops and ideas, and the installation of new plant before producing a commercial design at a price compatible with wood'.[25]

The proposed purchase, therefore, was hardly attractive from the manufacturing standpoint. When the question of profitability was addressed, however, an entirely different picture emerged (see Table 3). Supermarine was, in fact, extremely profitable, although the company's receipts in this period were no greater than those of the majority of airframe firms. 'The position of Supermarine is typical of all aviation businesses', Vickers Aviation's managing director, Sir Robert McLean, observed. 'They rely upon designing acceptable new types for the Air Ministry and when they succeed the profits are very large; when they fail it is still the policy of the Air Ministry to keep them alive.' Vickers' report noted that 'the average capital employed is small as the business has been conducted very largely with borrowed money'. The issued share capital was increased from £13,500 to £250,000 in 1927, but of the additional capital only £1 was received in cash. £236,500 was provided by the capitalisation of undrawn profits and by 'writing up' new fixed assets which had been purchased on exceptionally favourable terms.[26]

Vickers was convinced that Supermarine's profitability could be maintained or even increased. The change from wooden to metal construction would allow the work-force to be diluted and labour costs to be cut. Production expenses could be further reduced by simplifying design. There was some scope for staff reductions and for co-operation between the Vickers and Supermarine design teams. 'The two organisations', McLean wrote, 'should supply us with a strong and progressive design staff, and we should be in a position to compete in all aircraft markets open to British machines.' Plans were ready for a three-engined flying boat to replace Supermarine's existing model. 'If Supermarine succeeds the business is assured of handsome profits for some seven years.'[27] In fact Supermarine's average turnover in the five years prior to rearmament was 35 per cent higher than between 1923 and 1927.[28] Percentage profits were reduced by competitive tendering for subcontracts, but this encouraged the company to improve its efficiency. In December 1931, for example, Supermarine reported that 'although there is a substantial reduction in the number employed in the works, improved production facilities have enabled us to increase . . . the output of work for 1931 as compared with the previous year'.[29] Vickers' turnover also increased: in the five years prior to rearmament annual sales averaged £580,000.[30] Relatively high levels of output and investment enabled the company to employ a very much larger proportion of unskilled workers than many other aircraft firms and to achieve a significant reduction in manufacturing costs.[31]

Vickers and Supermarine sometimes pooled their design resources and were not jointly managed for financial reasons alone. Hence Vickers' Wellington bomber, which was produced in larger numbers than any other British bomber during the Second World War, benefited from design experience gained from both Supermarine's Stranraer flying boat and Vickers' Wellesley general purpose aircraft.[32] The firms were also supplied with ample funding for research and development by their holding company. A grant of £250,000 was provided for this purpose in 1929 after which, McLean later recalled, there were 'no financial worries at all'.[33]

The formation of Hawker Siddeley

For Vickers, aircraft manufacturing in the inter-war years was a sufficiently attractive commercial proposition to justify expansion. However, profit played a rather different part in the formation of the Hawker Siddeley group in the early 1930s. Hawker's position in the previous decade had been somewhat unstable but the subsequent success of the company's designs led to a rapid revival in its fortunes. Turnover reached £1 million in the year ending March 1935.[34] Profits increased in every year between 1931 and 1937 and during the early 1930s averaged as much as 28 per cent on Air Ministry sales,[35] yet with so many different firms in the airframe industry it is hardly surprising that Hawker also resorted to other tactics to overcome competition.

In the last months of 1930 it became clear that Air Ministry requirements for the Hawker Hart bomber during the following year would overburden Hawker at the expense of several other contractors. The Ministry therefore proposed to invite competitive tenders for the available contracts. The industry had consistently opposed this policy during the late 1920s and in November 1930 the SBAC approached the Member of the Air Council for Supply and Research (AMSR) seeking a conference on the general question of subcontracting, the term used by the SBAC for competitive tendering. Richard Fairey, of Fairey Aviation (Hawker's foremost competitor) suggested that the Air Ministry should order two types from each operational class in order to maintain design staffs. Frederick Handley Page likewise argued that technical development would only progress if firms were allowed to produce their own designs. The Air Ministry, however, had no illusions about where the real pressure against competitive tendering was coming from. On 27 March 1931 the Air Council minutes noted that 'Messrs. Hawkers . . . had induced the Society of British Aircraft Constructors to support them by making representations against the whole principle of subcontracting'.[36]

The conference took place later that month with Fairey, Sir Robert McLean of Vickers, and Thomas Sopwith of Hawker in attendance. The industrialists presented a united front against competitive tendering, denied that it was an economic method of purchase, and suggested that the Air Ministry should either order two designs from each class or place the responsibility for subcontracting with the designing firm. The Air Ministry was, however, under pressure from Parliament to reduce prices through competition and in the following month it was decided to invite tenders for the Hart.[37]

The apparent solidarity within the SBAC crumbled immediately. A price of £3160 per aircraft had been paid to Hawker for the first Hart production contract; the firm's lowest quotation was £2300. Vickers now tendered for £1800 and won the contract. In July, Sopwith was still arguing 'that his firm employed a considerably larger proportion of fully skilled workers than Messrs. Vickers [and] could not contemplate making "Harts" at the Vickers price'.[38] But subsequent contracts undertaken by Vickers and Armstrong Whitworth saw the price fall still further to £1475. In May 1933, the Secretary of the Air Ministry, Sir Christopher Bullock, recorded that 'the result of our firm insistence on contracting out on that occasion has been, directly and indirectly, to save the public purse a sum running into something like £1 million . . . Though Messrs Hawker must have full credit for a high degree of technical competence . . . I consider that they have been amongst the worst profiteers in the industry.'[39]

There can be no certainty that Hawker's subsequent actions were motivated purely by the price-fixing disputes of the early 1930s. The formation of the Hawker Siddeley group occurred at a time of escalating demand and of unprecedented enthusiasm for aviation in financial and industrial circles. Moreover, the decline of former market leaders like Fairey and Napier served as a warning to the more successful firms that the Air Ministry could always switch its patronage to other companies and one means of preventing this was to buy the other companies out. Nevertheless, it was primarily the SBAC's failure to protect their financial interests in 1931, together with further disagreements with the Air Ministry in 1933, which persuaded the Hawker directors that their trade organisation would never bargain effectively while the industry remained fragmented.

Commercial success allowed Hawker to expand its operations, however. In 1933, a new public company, Hawker Aircraft Ltd, was established with an issued capital of £787,000. In the following year Hawker purchased Gloster Aircraft. In January 1935 issued capital was increased by £100,000 and, in July, the formation of Hawker Siddeley Aircraft brought together the two Hawker firms and the entire Armstrong Siddeley Development Company.[40] The next logical step was a merger with Vickers and Supermarine; seven of the industry's most important firms would then have been con-

trolled by one holding company. An approach was made at the end
of 1935, and while the idea of outright purchase was instantly dismissed, consideration was given to a co-operative agreement under
the terms of which several Hawker Siddeley directors, including the
managing director, Frank Spriggs, would have joined the boards of
Vickers Aviation and Supermarine.

Some Vickers directors may have been attracted by this idea but it
was unacceptable to Sir Robert McLean. 'The Hawker Siddeley
directors who would come on to the board were absentee directors
plus an accountant with small appreciation of the technical
problems involved', he wrote. 'The burden of management would
fall largely on Vickers . . . It had been represented to Vickers . . .
that the resulting group would carry greater weight with the Air
Ministry', but McLean felt that 'the Air Ministry would take the
view that so powerful a group would be able to grow through lean
years without Air Ministry help, and that the Air Ministry would
tend to abandon it to its fate.'[41]

Vickers' rejection and, perhaps, the impracticability of competitive tendering after 1935 ended Hawker Siddeley's quest for
further amalgamations in the pre-war years. Nevertheless, a major
new company had been created involving, in Hawker and A.V. Roe,
two of the most experienced British aircraft constructors.

The consolidation of aero-engine production

A tendency towards consolidation can also be seen in the aero-
engine industry in the years prior to rearmament. Military aero-
engine manufacturing in inter-war Britain was largely undertaken
by four firms: Armstrong Siddeley, Napier, Rolls-Royce, and
Bristol. After fluctuating in the mid-1920s, Armstrong Siddeley's
aero-engine sales remained relatively stable until the beginning of
rearmament.[42] A market was maintained by airframe firms within
the Armstrong Siddeley group like A.V. Roe, who designed the
successful Tutor trainer around an Armstrong Siddeley power
plant. But Armstrong Siddeley failed to make a strong enough
commitment to research and development and its engines were
rarely selected for first-line aircraft during the 1930s.[43]

The same mistake was made by Napier. Napier's Lion engine
dominated the military market until 1930: Air Ministry payments to
the company from 1923 to 1930 totalled £5.4 million, which was
more than twice the sums paid to Bristol, Armstrong Siddeley and

Rolls-Royce; in the same period annual profits averaged £190,000, and gilt-edged investments totalling almost £1 million had been accumulated by 1927; the company was, quite literally, embarrassed by the scale of its earnings. Yet it was unable to find a successor to the Lion, while competitors like Bristol and Rolls-Royce were engaged in the development work which eventually produced engines like the Pegasus and the Kestrel. Consequently, Napier's position deteriorated rapidly during the early 1930s. Having declared higher profits than any other public aircraft company in 1929, earnings were halved between 1930 and 1931, and in 1933 Napier made a loss.[44]

The decline of Armstrong Siddeley and Napier had one positive side-effect in that it illustrated the importance of an aggressive approach to research and development. When, at the height of its prosperity, Napier was invited by the Air Ministry to develop a new engine similar to the American Curtis D-12, the company refused and Rolls-Royce took on the work instead. The result was the Kestrel engine, which established Rolls-Royce as one of Britain's foremost military aero-engine manufacturers in the early 1930s. Between 1930 and 1934 Rolls-Royce produced a total of 1882 engines. Gross aero-engine profits in the same period rose from £177,000 to £597,000; they may well have been used to subsidise the company's motor car division.[45]

Rolls-Royce was overwhelmingly dependent on the military market. Bristol, on the other hand, was both a military and civil manufacturer. The company's Jupiter model was one of the most widely used aero-engines in the world in the later 1920s; 7100 were eventually produced. Between 1921 and 1928 eight times as many Bristol-designed engines were produced in Europe as any other design. Yet contemporary trading practices ensured that these impressive figures were not directly reflected in the scale of Bristol's own manufacturing operations for many of their designs were pro-duced abroad under licence. Licences to manufacture the Jupiter were, for example, sold by Bristol to no fewer than 17 different countries. At the end of 1932 the Jupiter was succeeded by the Pegasus and the Mercury and, during the next two years, 1000 of these engines were produced, many of which were exported.[46]

By the onset of rearmament, Napier had virtually no work and the future of the company had become a matter of serious concern to the Air Ministry.[47] Nearly all RAF first-line aircraft were powered by Rolls-Royce and Bristol engines and both firms were developing new engines like the Merlin, Perseus and Taurus, which eventually entered service in the later 1930s.

Consolidation in the civil market: De Havilland

One consequence of the RAF's control over the bulk of the inter-war aircraft market (and the resulting emphasis on military aircraft design) was that many British contractors neglected an important alternative source of demand – the private flyer. The military market therefore remained divided between a relatively large number of contractors, while the private market became the virtual monopoly of De Havilland. The maximum production of any single De Havilland model during the early 1920s was just 17, but the introduction of the 'Moth' for the private flying market resulted in a dramatic change in the company's fortunes. A total of 1762 Moths were built in Britain alone. Production on this scale provided De Havilland with the opportunity to introduce far more efficient manufacturing methods than those previously employed. The first fuselage jigs used for Moth production cut assembly time from one week to just one day.[48]

The aero-engines then available in Britain were generally too large and too expensive in terms of price and operation to be suitable for private aeroplanes, so De Havilland began designing and manufacturing cheaper small-capacity engines for installation into their aircraft.[49] To fund the expansion of business, capital was increased from £50,000 in 1924 to £400,000 in 1929, and more than £155,000 was spent on new buildings and on plant and machinery which included high-precision machine tools for aero-engine work.[50] Yet despite De Havilland's successful exploitation of the civil market and the enormous international demand for the company's products, average annual sales between 1930 and 1934 only amounted to approximately £500,000 (see Table 5), which may be compared to Vickers Aviation's annual average of £580,000 in the same period from a 100 per cent military (and predominantly British) market.[51] In fact, like Bristol, De Havilland's total production was limited by the common practice of licensing aircraft manufacture to foreign companies. Both France and Norway were granted licences to build De Havilland products and an agreement was made with the Wright Corporation for the production of De Havilland engines in the United States in 1928. Subsidiary companies, such as De Havilland Australia and the Moth Aircraft Corporation of America, were also established in the later 1920s. Total UK production of De Havilland civil aircraft numbered at least 3200 from the commencement of Moth deliveries to the outbreak of war in 1939, but the numbers would have been very much greater had so many of the company's aircraft not been built abroad.[52]

TABLE 5

De Havilland Sales (£), 1924–35

Year	Sales	Year	Sales
1924	138,495	1930	543,888
1925	187,060	1931	487,141
1926	222,902	1932	376,588
1927	299,022	1933	523,078
1928	387,102	1934	607,713
1929	676,529	1935	1,018,318

Source: HSA, De Havilland acounts.

During the Second World War, responsibility for producing the key operational aircraft designs was shared by the professional aircraft firms and engineering companies from the electrical and automobile industries. However, the most important professional firms, measured by their total wartime output, were Vickers, Hawker Siddeley, Rolls-Royce, Bristol and De Havilland. At the beginning of rearmament these firms already dominated the market. Judged by the standards of aircraft companies in other countries, they were large concerns. They employed a total of 22,000 people (66 per cent of British aircraft employment) in September 1935,[53] and produced the vast majority of British aircraft and aero-engines very profitably.

Correlli Barnett's study, *The Audit of War*, suggests that the task of pre-war and wartime expansion would have been easier had these firms gained more production experience prior to rearmament.[54] This can only be a matter for conjecture, but the implicit assumption that the health of the inter-war aircraft industry can be measured by the volume of its operations is misleading. British aircraft production increased steadily between the mid-1920s and 1935, when Britain actually produced more aeroplanes than the United States.[55] In so far as the number of aircraft imported into Britain was negligible, the only scope for any further expansion of demand lay in the export market. Yet rather than exporting aircraft, British firms like Bristol and De Havilland often preferred to licence their products to foreign manufacturers, and licensing agreements were sometimes required by foreign governments as a condition of purchase.[56]

In short, the foreign demand for British aircraft did not necessarily translate itself into production contracts for British aircraft companies. Given the market open to the industry and the trading practices of the period, it is doubtful whether even the largest firms could have grown much further before the commencement of rearmament. In any case, the emergence of these same companies as the key wartime contractors suggests that they were not unprepared by qualification or experience for the problems of war production.

THE AIR MINISTRY AND MILITARY AIRCRAFT PRODUCTION

The adequacy of overall capacity for both peacetime and wartime production was, of course, chiefly the concern of the Air Ministry rather than the aircraft firms. The Air Ministry's views on the strength of the peacetime industry changed as the trend towards consolidation became more pronounced in the early 1930s, but in the immediate aftermath of the First World War the industry's problems caused considerable disquiet in official circles.

In 1918 British aircraft production had averaged 2500 per month, but the post-war contraction of the industry greatly reduced its capacity. By 1924 the Air Ministry's Department of Supply and Research was estimating that the industry might produce just 55 aircraft in the first month of a war. Total requirements for the first year of hostilities were estimated at 25,000 but, although this number had been exceeded in 1918, it was thought that twice as much effort would now be needed to achieve the same result because of the increasing complexity of the aircraft in service.[57]

The shortfall was even greater in the aero-engine sector. In 1918, 3.9 million horsepower had been produced by the aero-engine industry. By 1925 the Air Ministry's requirements for the first year of a war were equivalent to 15 million horsepower and the responsible director of aircraft inspection was warning that 'the steady increase in output per engine unit which has persisted since the war period . . . has most seriously reduced the possibilities of sudden expansion'. It was thought that Rolls-Royce, one of the largest wartime suppliers, might produce the equivalent of one million horsepower in the first year of a war.[58]

When the Air Ministry realised the extent of this deficiency a lengthy inquiry was held into the whole issue of war production. The investigation took place in the aftermath of the so-called 'demobilisational instability' period, and detailed attention was given to the difficulties facing the firms at that time. It soon became clear that the effectiveness of wartime expansion would be critically influenced by the health of the peacetime industry. In an interim report to the AMSR, the Air Ministry's Principal Supply Officer (PSO) argued that there were too many firms and too many different types of aircraft:

It is a matter for consideration . . . whether the available money . . . is better spent by spreading contracts over as great a number of firms as possible, or whether the better policy would not be to concentrate on a smaller number of those approved and tried in the past . . . This alternative would at any rate have the effect of improving the financial status of those selected, of enabling them to retain their staffs and operatives on a more permanent basis, and of causing a greater evenness of pressure in output.[59]

It was also recognised that official procurement procedures often discouraged the industry from employing efficient production methods. New types of aircraft were frequently ordered before they had passed out of the experimental stage and were then constantly modified while in service; deliveries were called for which were either impossible or, if not impossible, sufficiently urgent to prevent production from being organised scientifically; and the annual defence estimates were prepared in a manner that resulted in peaks and depressions in the work which the firms had to do, which in turn increased production costs. The Air Ministry's supply staff argued that a longer-term production programme was needed:

Without this careful planning any attempt to approximate to mass production methods with an eventual saving in time and cost, or to plan for war production with a certainty of requirements being forthcoming, will be defeated.[60]

There were also complaints about the Air Ministry's research and development organisation. The Ministry's Directorate of Technical Development (DTD) was accused of 'fostering its activities at the expense of the squadrons in the matter of their equipment'.[61] It was argued that the standard of construction and the specifications of material demanded by the Air Ministry were too high, and that contractors were required to install an excessive variety of different equipment.[62] By January 1926 the PSO had concluded that the aircraft industry in itself, with every circumstance in its favour, would fall short of requirements by at least 60 per cent in the case of aircraft for the major emergency and by, at least, 80 per cent in the case of engines.[63]

Responsibility for aircraft research, development and production rested with the AMSR, Air Chief Marshal Sir Geoffrey Salmond. Salmond was more sympathetic to the DTD than were his supply staff and he doubted the wisdom of concentrating Air Ministry orders on a small number of proven manufacturers, suggesting that the ministry would 'be in danger of being faced by a trade combine as regards prices'.[64] In both 1924 and 1925 he therefore rejected proposals to reduce the number of Air Ministry contractors. He did not initially see any scope for implementing other radical reforms in procurement procedure either. The immediate pressure to re-equip the squadrons was, he argued, too great, although he hoped for more stability in the long term.[65]

By the following year, however, so much attention had been drawn to the unsatisfactory position of the aircraft industry and to its dangerous implications for war production that Salmond became convinced that drastic action was necessary. The PSO was recommending a complete revision of procurement policy. A five-year re-equipment programme would be adopted for each class of aircraft; only one design for each of the ten service classes would be ordered; the existing number of military airframe contractors would be reduced to twelve; and those firms whose designs were selected would become 'master production firms' for each of the ten classes for the next five years.[66] Salmond gave his full support to these proposals. In passing them to the Chief of the Air Staff (CAS), Air Chief Marshal Sir Hugh Trenchard, Salmond pointed out that they dealt with the whole problem economically by providing a regular flow of production orders to the aircraft firms, thus maintaining those firms in a stable financial position. The many problems caused by the large number of designs in service would be eliminated and the cost per aircraft would be reduced by as much as a third.[67]

Believing that war production would be accompanied by insuperable problems unless the aircraft industry was placed on a sounder peacetime footing, Salmond expected immediate action to be taken in response to his recommendations. He reckoned without Trenchard's opposition. Trenchard understood war production to be a problem which would have to be solved in wartime. He agreed in principle with the PSO's scheme, but considered it too ambitious to be practicable in the immediate future. Action was therefore limited to reducing the number of designs, testing aircraft in squadrons for a year before placing larger orders, and retaining aircraft in service for seven years.[68]

It is difficult to know how well-founded the PSO's anxiety about the aircraft industry really was, and how effective his recommendations for rationalising procurement would have been. His proposal to reduce the number of Air Ministry airframe contractors to twelve would have had little effect for between 1924 and 1930 only nine companies received regular production contracts in any case, and three of the Ministry's 14 design firms undertook virtually no production work at all. Moreover, even if the volume of output of the more successful firms was limited by wartime standards, the industry was experienced in the problems of large-scale production, for throughout the inter-war years it remained under the control of many of the same senior executives, managers and technicians who had presided over expansion during the First World War.[69] Equally questionable is the implication that a more efficient industry might alone have satisfied the wartime demand for aircraft; it is certain that additional capacity would have been required under any circumstances.

However, the problem of wartime expansion might have received much less attention in government circles in subsequent years had the PSO painted a more sanguine picture of the industry's position. In the event, his warnings alerted the Air Ministry to the need to devise a war production plan to bridge the gap between peacetime capacity and wartime requirements.[70] The plan was based on three broad principles: conversion, concentration and organisation.

The principle of conversion involved harnessing industrial resources from outside the aircraft industry to the manufacture of aircraft and aircraft components in war. The most promising source of supply was the engineering industry, particularly the motor car industry, but it would have been impossible to exploit this while service aircraft were still constructed from wood, so the decision was taken to order metal aircraft instead.[71]

The principle of concentration involved focusing as many industrial resources as possible on a small number of proven aircraft. The aim was to maximise efficiency by achieving economies of scale. It was envisaged that, on the outbreak of war, aircraft and motor car firms would be organised into groups; one group would be created for each of the most important designs, and only one design would be selected for each operational category.[72]

The third principle, organisation, involved the creation of a war production organisation within the Air Ministry to plan and co-ordinate the supply of materials and such equipment as engines, instruments, armament and wireless apparatus, which the Ministry normally purchased and supplied free as an 'embodiment loan' to the airframe firms. The firms themselves were expected to form committees to co-ordinate the purchase and distribution of materials within their respective groups.[73]

So although a long-term re-equipment programme was not introduced, the PSO's investigation of the aircraft industry's capacity did have some very positive consequences. A wartime mobilisation plan was drawn up; a general transition from wood to metal construction took place in the following years, and, by subcontracting the all-metal Armstrong Whitworth Siskin fighter to four different firms, the Air Ministry also succeeded in limiting the number of designs in service. Orders were also placed less sporadically during the later 1920s by combining total demands for the financial year into single contracts.[74]

Rationalisation was rejected as an object of policy. Hence an offer from one firm (almost certainly Vickers) to produce all the Siskins at a price of £1650 each was refused and the Air Ministry spread the orders over four companies at a higher cost per aircraft, partly to keep them in business and partly to educate them in the use of metal construction.[75] Nevertheless, a relatively small number of contractors established themselves as market leaders in the following years and when, in 1931, a specially appointed Aircraft Supply Committee undertook the next major enquiry into aircraft production and the condition of the industry, there was more scope for optimism:

The aircraft industry is sturdier and more healthy than it was ten years ago. Aircraft firms appear to have been almost alone in withstanding the effect of the general industrial depression. Capital is now securely rested; consolidations have brought strength; foreign trade is beginning to appear. There is no longer a general need for artificial shelter to the same extent as hitherto, lest firms should be driven out of business in such numbers as to endanger the nucleus required for war.[76]

The committee also noted that orders were increasingly being concentrated with the more successful design firms. It was pointed out, however, that this development had brought several companies to the verge of liquidation and it was thought likely that the number of Air Ministry contractors would have to be reduced.[77]

If the Supply Committee recognised that the leading aircraft firms were now financially stronger than had been the case in the mid-1920s it also acknowledged that the Air Ministry's procurement policies were still impeding the development of improved production methods.[78] The importance of the Committee's findings, however, lies less in the production problems it described than in its clear understanding of those problems. Neither ignorance nor conservatism on the part of the Ministry or the aircraft industry was preventing the use of better manufacturing techniques. What was needed was a change in the way aircraft orders were placed.

One possible solution was to give the firms longer production programmes, as the PSO had suggested in 1926. The earlier proposal for a five-year re-equipment scheme was considered too ambitious, but it was felt that three-year manufacturing programmes might bring many of the same advantages.[79]

Two main difficulties were involved in introducing such programmes. First, the scope for competitive tendering would have been reduced and the Air Ministry would consequently have had to forfeit much of its control over aircraft prices. It was, naturally enough, extremely reluctant to do so. In 1931 the Hawker Hart was in the process of becoming the standard RAF bomber and, on this basis, might well have been selected for a three-year programme, but instead, as already noted, the Air Ministry decided to invite competitive tenders for the aircraft. Prices fell dramatically as a result. It is most unlikely that the allocation of a three-year Hart contract to Hawker would have reduced prices to anything like the same extent.

Second, three-year programmes necessarily entailed a sacrifice of technical quality in return for quantity.[80] For some types of aircraft, such as training aircraft, however, such sacrifices could be accepted, and one contractor, A.V. Roe, was therefore given an extended programme for the Tutor trainer, starting in 1933 and concluding in 1936.[81] The aircraft was the first to be accepted by the RAF with an all-welded fuselage and empennage, which was far cheaper in production than riveting and bolting or the traditional strut and socket biplane structure.[82] The three-year programme enabled A.V. Roe to develop new mass-production methods involving extensive jigging and tooling. A total of 795 Tutors was eventually built for the RAF and overseas air forces.[83] It is significant that by 1936 A.V. Roe was being described by the Air Ministry as 'the best organised manufacturing unit we have [with] extremely efficient and well-equipped tool room, drawing office and machine shops'.[84]

The Air Ministry hoped to raise industrial efficiency through the allocation of three-year programmes in the belief that this would prepare the aircraft firms for expansion in an emergency. Yet even if the Ministry's efforts had been completely successful there would still have been a substantial gulf between peacetime capacity and wartime demand, and it was still expected that any deficiency would be made up by drafting automobile and other general engineering factories into the wartime aircraft economy. By this time such matters were the responsibility of the Principal Supply Officers' Committee of the Committee of Imperial Defence (PSOC).

The proposals made by the Air Ministry in 1926 for wartime production groups of aircraft and engineering firms had been passed to the PSOC and during the following years it made detailed assessments of national manufacturing potential and formulated a series of plans for its allocation to the supply ministries in an emergency.[85] These plans were not merely hypothetical. In 1931 Daimler agreed to make its facilities available for aero-engine production in wartime and, consequently, the company was given a large 'educational' subcontract for Bristol engines.[86] In December 1932 the PSOC decided that four motor firms, Standard, Daimler, Humber and Singer, would be allocated to the airframe sector in wartime, and a year later similar plans were approved for aero-engine production. Production from the aero-engine industry would be supplemented by the machining capacity of eight motor firms, five of which would produce parts and sub-assemblies, and three of which would build complete engines. Bristol actually drew up plans for the production of 5800 engines involving the entire machining capacity of Daimler.[87]

The plans, however, were based on RAF requirements for overseas defence alone during the early 1930s. These amounted to 7444 aircraft in the first year of hostilities. It was concluded, therefore, that there was only a comparatively narrow gulf between industrial capacity and service needs, and that a surplus would be available after 12 months.[88] Further analysis of war requirements was undertaken during rearmament and will be considered in a later chapter. It is sufficient to state here that these calculations (which included home defence) initially envisaged the production of 15,000 aircraft in the first year of a war, and later of 20,000; and as requirements increased so, too, did estimates of the likely deficiency, influenced by the production problems which the industry encountered in the early stages of rearmament.[89]

AIRCRAFT DESIGN AND CONSTRUCTION

If the production techniques employed by the aircraft industry were critically influenced by the Air Ministry's procurement policies, so too was the pace and direction of aeronautical technology in inter-war Britain. Although the military selection process was flexible enough to exploit so-called 'private venture' designs proposed by the industry, the aircraft purchased by the Air Ministry between the wars were usually designed to official specifications. Moreover, the state played a major part in advancing aircraft design through research bodies like the National Physical Laboratory and the Royal Aircraft Establishment, and through direct contracts with the aircraft firms. According to one contemporary estimate the government spent more than £15 million on aeronautical research and development between 1925 and 1934.[90] In addition to contributing specifications, funds and facilities for research and development, the government also provided the main market for aircraft and, as the product itself had to be fashioned to the needs of its principal purchaser, military requirements tended to shape the progress of design and development.

One firm, De Havilland, prospered without Air Ministry patronage by conquering the market for cheap, economical, low-performance private aircraft. Others found it much more difficult to locate alternative markets for their designs. Both Handley Page and Armstrong Whitworth lost heavily on civil contracts with Imperial Airways.[91] On eight different occasions between 1922 and 1935, Blackburn resolved to reduce its dependence on the RAF and to enter the civil aircraft market but, reviewing this policy in 1935, the company's chairman was forced to admit that neither its civil aircraft nor its flying boats had 'achieved any measure of success, and . . . little or no profits had accrued from these programmes' despite the expenditure of 'considerable sums of money and time in pursuing their development'. Blackburn concluded: 'There seems no doubt that the military side, on which we are largely dependent, had so predominated as to throttle the other two programmes.'[92]

This view was echoed by other leading manufacturers. Richard Fairey, chairman of the most successful military aircraft company of the 1920s, argued in 1928 that military aircraft construction required so much concentration of effort that civil types could not be produced as well.[93] Sir Robert McLean of Vickers similarly doubted the capacity of his company to design civil and military prototypes. He advised the Vickers board against the diversion of design staff 'from the immediate and profitable market' for military aircraft. 'I am anxious', he wrote, 'not to commence broadening our base by . . . digging away some of our existing foundations, which are the Air Ministry.'[94]

It is hardly surprising, then, that the Air Ministry was directly responsible for two of the most important technical advances of the later 1920s. One, the development of more powerful liquid-cooled engines for military aircraft by Rolls-Royce, resulted from the Air Ministry's desire to see a British manufacturer develop a 'monoblock' engine like the Curtis D-12.[95] The success of this project gave Britain an important lead in the development of small capacity, high-power, engines which eventually proved particularly well suited to interceptor fighters like the Spitfire and, later, the American Mustang. The other, the redesign of the standard wooden biplane in metal, was sponsored by the Ministry to increase the supply of aircraft in an emergency.

Several different forms of metal construction were developed in inter-war Britain and were subsequently incorporated into the monoplanes of the rearmament and wartime period. This lack of standardisation can be traced back to the later 1920s. In December 1925, for example, the board of Blackburn Aircraft considered whether it should concentrate resources on 'one particular kind [of metal construction]' or exploit 'various methods and systems'. The firm eventually decided it would 'exploit various systems of metal construction and not tie [itself] to one type only'. Three types were under consideration: 'Pure dural [aluminium alloy stressed-skin] on the lines of Junkers. High grade steel strips as Boulton and Paul. Mild steel which Blackburn and Hawker's were developing.'[96]

A similar situation prevailed throughout the industry. Reporting in 1926 on the progress of the transition to metal, the Air Ministry's Director of Contracts professed himself unsure 'whether technical opinion is sufficiently hardened to enable a choice to be made between the sheet method (Junkers), the tube type (Hawker and Fokker) and the strip type'. Another report, written three years later, again recorded that 'no one metal has been standardised . . . At present we have three parallel systems of steel strip, duralumin, and steel tubes.' In the following year Vickers commenced research on a fourth method – geodetic construction.[97]

The improvement of aero-engine performance and the development of metal construction were of great military value yet both tended to divert technical resources from aerodynamic research and development which, in the United States, led to the replacement of the biplane by the monoplane. Hence the Rolls-Royce Kestrel engine brought substantial advances in performance to aircraft of traditional biplane configuration largely by increasing the ratio of engine power to airframe weight. Metal construction was accompanied by many advantages in terms of cost, strength, maintenance and storage, and was desirable from the war production standpoint. It did, however, require a much greater technical effort than wooden construction and did not in itself improve performance.[98]

For some years this was unimportant because speed was not the only criterion for military selection, and biplanes often fulfilled the RAF's demands more effectively than monoplanes.[99] The British aircraft industry was capable of designing monoplanes of very great aerodynamic efficiency, like the Supermarine S.6, which won the Schneider Trophy race in 1929. The S.6, however, was not sufficiently manoeuvrable to be converted into a fighter and it was estimated by one leading industrialist that other changes that were needed to turn it into a military landplane would reduce its top speed from 350 to 260 m.p.h. As the RAF's biplanes were already attaining speeds of more than 200 m.p.h. with less than half the engine horsepower of the S.6, he argued that the cost of any such conversion was impossible to justify.[100]

This line of thinking was broadly correct, for some of the early military monoplanes, like the Boeing P-26 or the Dewoitine D-510 fighters, were not significantly faster than the last generation of biplane fighters, such as the Gloster Gauntlet or the Fiat CR-32. Biplanes therefore remained in service in the British, German, Italian and Czechoslovakian air forces for much of the inter-war period.[101] The monoplane's potential advantage in terms of speed, however, became increasingly clear during the early 1930s, causing the RAF and its contractors to revise many previous assumptions about military design. Vickers commenced work on the first geodetic monoplane in April 1932, and Hawker's first proposals for a 'Fury monoplane' (which evolved into the Hurricane) were made to the Air Ministry in August 1933.[102]

American manufacturers took a different course. In the United

States the most significant technical advances were achieved not in the military sphere but in the design of large civil transport land-planes. Intense competition between the airlines created a demand for high-performance aircraft with low operating costs. This alone hastened the development of monoplane airliners.[103] Such aircraft, however, had to be built at a price that the airlines could afford to pay, and the economics of production also tended to favour the monoplane. For there were two ways to improve performance. One, the reduction of weight, was generally achieved through the design of relatively complex internal aircraft structures which were expensive to produce. The other, the improvement of external aerodynamic efficiency, required a greater technical effort but allowed firms to design simpler, if heavier, aircraft which were less costly in production.

This was the practice adopted by American designers, whereas British efforts continued to focus on the improvement of power-to-weight ratios.[104] The first all-metal stressed-skin civil monoplanes entered service in America in 1933, and by the following year the same technology was being incorporated into military aircraft, while in Britain the first monoplanes were still on the drawing board.[105] In November 1934 the AMSR, Air Marshal Sir Hugh Dowding, met senior executives from 14 aircraft firms and drew attention to Britain's 'backwardness in comparison with the very great progress which has been made recently by some other nations'. By this time he was well aware that several monoplanes were in the course of design and development. The difficulty was that it took 'much too long to get an aeroplane through from the design stage to the production stage'.[106]

The Ministry's own analysis suggested that there were several reasons for this. First, Air Staff specifications often made too many different demands. The DTD 'found Air Staff requirements indifferently stated and dotted about all over the specification'. Secondly, the number of experimental contracts in progress at any one time placed an excessive burden on the industry's designing capacity.[107] This view is certainly corroborated by the records of Blackburn Aircraft, which list numerous transient projects that foundered during the early stages of their conception, but even the most successful firm of the early 1930s 'tendered for everything'.[108]

However, the fundamental problem was best described by the government's industrial adviser, Lord Weir, in the following year. He wrote that

The outstanding feature of the existing system is a sort of evasion of responsibility by a procedure which postpones decision until the choice of type becomes almost automatic. It may result in a reliable decision, but the march of progress renders the ultimate selection obsolete.[109]

Of a maximum development schedule of eight years for a new bomber, an inordinate amount of time was absorbed by official procedures. Out of the 96-month timetable no fewer than 43 months were accounted for by such procedures, including 13 months of tests at Martlesham Heath and a 12-month development period during which alterations were incorporated and production requirements discussed.[110]

Reviewing the problems of aircraft design, development, and production in 1934, the Air Ministry concluded that Dowding should be freed from responsibility for production and allowed to concentrate on research and development during what was clearly a critical stage in the evolution of aeronautical technology. In December 1934, a separate Department of Research and Development was therefore created and Dowding's supply directorates were transferred to the Member of the Air Council for Supply and Organisation (AMSO), Air Chief Marshal Sir Cyril Newall.[111]

Then, in February 1935, the SBAC recommended that specifications should be simplified and that the industry should 'be relieved from DTD's control during the process of design and construction of a prototype'. As similar proposals had already been made within the Air Ministry, Dowding now agreed 'that the responsibility for all matters affecting design . . . should be placed on the designer with a minimum of intervention by the Department'. The 12-month development period was abolished altogether.[112]

The strengths and weaknesses of these reforms would only be demonstrated in subsequent years but much depended on the ability of the aircraft industry to absorb the new technology rapidly. The choice between developing streamlined monoplane structures *de novo* and exploiting breakthroughs already achieved in America was really no choice at all, and during the early 1930s British firms dispatched numerous technical missions to the United States to gain first-hand knowledge of the progress which had been achieved there.[113] The extent to which British design and development was both influenced and accelerated by these missions can be gauged from the Vickers' records. When, in March 1934, Vickers sent a technical team to America, the company had been engaged in the design of a particular bomber for two years. However, after the return of the mission 'all work done was scrapped and a new start was made on [the aircraft which was to become] the Wellington.' Production commenced just four years later, or in half the time the Air Ministry thought was required to design and develop a new bomber. The initial 1932 specification had envisaged a bomb load of 1000 lb and a range of 720 miles. The Wellington's bomb load was 4500 lb, and its range was 1200 miles.[114]

Yet the different evolutionary paths taken by the British and American aircraft industries ensured that British aeroplanes retained many distinctive and original features. As late as 1939, when most American and German companies were building all-metal aircraft, De Havilland chose to build the Mosquito out of wood.[115] It was probably the most successful fighter-bomber of the war. Hawker's Hurricane fighter, though equipped with metal-skinned wings in 1939, employed the tubular steel structure developed for the Hawker biplanes of the early 1930s.[116]

The best example is again provided by Vickers, which remained convinced of the potential of geodetic construction. The attraction of geodetics was entirely military. The technique produced aircraft which were light in relation to their all-metal equivalents and which therefore promised to improve the ratio of engine power to airframe weight. Geodetic aircraft were also exceptionally strong. At a time when there was a particularly pressing need to improve the performance of larger military aircraft, and when the Air Ministry's selection policy was overwhelmingly governed by qualitative considerations, these features alone justified the development of geodetics. That such structures were also expensive to produce was of secondary importance. The production problems had to be solved retrospectively through the standardisation of unit parts and the development of special machinery.[117]

CONCLUSION

Aircraft manufacturing in inter-war Britain has often been seen as an unprofitable business involving erratic orders for small batches of machines.[118] The Air Ministry's policy of dividing experimental contracts between 14 firms inevitably prevented some from gaining either production experience or commercial reward between 1918 and 1935. Other companies prospered, however, and produced aircraft and engines in large numbers by the standards of the day. The most successful manufacturers of the inter-war years went on to dominate Britain's aircraft economy in the Second World War. One of the salient features in the history of the inter-war industry was the emergence of Vickers, De Havilland, Hawker Siddeley, Bristol and Rolls-Royce as market leaders.

The relationship between the Air Ministry's ordering policy and the production techniques employed by the aircraft industry was clearly understood by both sides. It was recognised that better manufacturing methods existed than those often employed by British firms. But the aircraft industry had gained considerable experience of large-scale production during the First World War and it was perfectly capable of improving its production methods when given the opportunity or incentive to do so. The trial three-year programme allotted to A.V. Roe enabled the company to jig and tool much more extensively than had previously been possible. Competitive tendering also encouraged firms to cut their costs during the early 1930s.

Consolidation and the improvement of production methods increased the capacity of the leading firms to expand in an emergency. Nevertheless, official concern about production weaknesses in other quarters of the industry stimulated the development of a war production plan to convert general engineering capacity to aircraft production, to concentrate production on a small number of proven designs, and to create a production organisation within the Air Ministry. Quantitative problems were, therefore, expected, and preparations had been made to solve them.

The Air Ministry was much less concerned about the quality of British aircraft. Indeed, for much of the period the British genuinely believed that their military aircraft were technically superior to those of the other powers. This may well have been true until, with the appearance of all-metal stressed-skin monoplanes in the United States, performance alone began to outweigh all other military considerations. Drastic measures had then to be taken to catch up. At the onset of rearmament the aircraft firms were thus confronted not only by the problem of organising production on a scale unknown since the First World War but also of accommodating a new and revolutionary technology.

NOTES

1. P. Fearon, 'The Formative Years of the British Aircraft Industry, 1913–1924', *Business History Review*, 43 (1969), pp. 476–9; Edgerton, *Aeroplane*, 8–9, p. 14.

2. M.M. Postan, *British War Production* (London, 1952), pp. 4–5.

3. R. Higham, 'Quantity vs. Quality: The Impact of Changing Demand on the British Aircraft Industry, 1900–1960', *Business History Review*, 42 (1968), p. 445; HSA, Sopwith Aviation board minutes, 26 January – 7 September 1920; Sopwith accounts.

4. P. Fearon, 'Handley Page', p. 67.

5. Fearon, 'Formative Years', p. 493.

6. HSA, Sopwith board minutes, 14 October 1919.

7. P. Fearon, 'Airframe Industry', pp. 242–3; Higham, 'Quantity vs. Quality', pp. 447, 458. The principal beneficiaries of this system are listed in Table 2.

8. For example, of £2 million worth of aeronautical 'exports' recorded in the *Annual Statements of Trade of the UK* in 1930, £600,000 worth went to RAF squadrons stationed in India. See Vickers 322, E.C. Bowyer to the Council of the Society of British Aircraft Constructors, 28 January 1932.

9. P. Fearon, 'Aircraft Manufacturing', in N.K. Buxton and D.H. Aldcroft (eds), *British Industry Between the Wars: Instability and Industrial Development, 1919–39* (London, 1979), p. 231.

10. Fearon, 'Airframe Industry', p. 242.

11. PRO AIR 2/266, S.23447, Ffisk (Boulton Paul) to PSO, 5 February 1925.

12. Of those firms which left the aircraft industry in the 1920s only the English Electric Company succeeded in gaining re-entry during the Second World War.

13. Fearon, 'Airframe Industry', p. 240.

14. HSA, Blackburn board minutes, 11 July 1929; Blackburn Consolidated board minutes, 31 December 1931.

15. PRO AVIA 46/268, statement of value of aircraft contracts placed, 1924–41; PRO AVIA 46/72 statement of annual expenditure on aircraft and associated products, 1931–44.

16. *Annual Statements of Trade of the UK*.

17. AC 70/10/54, memorandum by C.V. Allen, solicitor of the Society of British Aircraft Constructors, 28 November 1935.

18. PRO AIR 2/619, C.R. Fairey (chairman and managing director of Fairey Aviation) to AMSR, 6 November 1930.

19. AC 70/10/9, Frederick Handley Page (managing director of Handley Page Ltd) to Sir Robert McLean (managing director of Vickers Aviation), 28 November 1933.

20. Weir 19/23, Lord Weir (the government's industrial adviser) to S of S, 15 July 1935.

21. HSA, Armstrong Whitworth board minutes, 12 December 1932, 30 November 1933; Armstrong Whitworth accounts.

22. HSA, Armstrong Whitworth board minutes, 29 March 1935.

23. Calculated from production figures in PRO AIR 19/524.

24. Vickers K757, memorandum by McLean, 29 October 1928.

25. Vickers K757, Muller to McLean, 25 September 1928.

26. Vickers K757, memorandum by McLean, 29 October 1928; memorandum by Messrs Deloitte, Plender and Griffiths (Vickers' accountants), August 1928.

27. Vickers K757, memorandum by McLean, 29 October 1928.

28. Vickers 61, statement of Supermarine turnover, 1930–34, submitted to the Royal Commission on the Private Manufacture of and Trading in Arms.

29. Vickers 163, Supermarine quarterly report to December 1931; Vickers 701, J.D. Scott's interview with T.C.L. Westbrook (former Supermarine production manager), 15 October 1959. Scott's interviews were conducted for his history of Vickers.

30. Vickers 61, statement of Vickers turnover, 1930–34, submitted to the Royal Commission on the Private Manufacture of and Trading in Arms.

31. Weir 19/23A, Air Ministry statement on employment in the aircraft industry, 30 September 1935; Vickers 169, Vickers quarterly report to 30 June 1933.

32. Weir 19/20, meeting between Weir and McLean, 6 February 1936.

33. Vickers 687, J.D. Scott's interviews with McLean, 8 and 9 November 1959.

34. HSA, Hawker Aircraft accounts.
35. *The Statist* (30 March 1935); Sir Christopher Bullock papers, the Secretary of the Air Ministry (Bullock) to US of S, 31 May 1933.
36. PRO AIR 2/619, Fairey to AMSR, 6 November 1930; Handley Page to AMSR, 29 December 1930; extract from Air Council minutes, 27 March 1931.
37. PRO AIR 2/619, US of S to Fairey, 2 April 1931.
38. PRO AVIA 8/158, meeting between the Air Ministry and Sopwith, 29 July 1931.
39. Bullock papers, the Secretary to US of S, 31 May 1933.
40. *The Statist* (30 March 1935 and 20 July 1935).
41. Vickers 687, McLean to Vickers Ltd board, 4 December 1935.
42. Armstrong Siddeley archive, Armstrong Siddeley accounts.
43. Fearon, 'Aircraft Manufacturing', p. 227.
44. C. Wilson and W. Reader, *Men and Machines: A History of D. Napier and Son Ltd, 1808–1958* (London, 1958), pp. 127–41; *The Statist* (30 March 1935).
45. I. Lloyd, *Rolls-Royce: The Years of Endeavour* (London, 1978), pp. 91–3, 97–8, 228, 232.
46. B. Gunston, *By Jupiter! The Life of Sir Roy Fedden* (London, 1978), pp. 48–55; AC 79/2, draft manuscript on the history of the Bristol Aeroplane Company, pp. 118, 129; 'company chronology', both by C.H. Barnes (no date).
47. PRO AIR 2/714, meeting between D of C and Sir Harold Snagge (managing director of Napier), 11 November 1935.
48. C.M. Sharp, *A History of De Havilland* (London, 1960), pp. 96, 400–7.
49. Ibid., pp. 98–103.
50. HSA, De Havilland accounts, 30 September 1924–1929.
51. Vickers 61, statement of Vickers turnover, 1930–34, submitted to the Royal Commission on the Private Manufacture of and Trading in Arms.
52. Sharp, *De Havilland*, pp. 111–13, 406–7. De Havilland also built large numbers of training aircraft for the RAF in this period.
53. Weir 19/23A, Air Ministry statement on employment in the aircraft industry, 30 September 1935. Comparative estimates of British, German, French and American aircraft employment indicate that the British aircraft industry was as large as any other until the mid-1930s. See Edgerton, *Aeroplane*, p. 26.
54. Barnett, *Audit*, pp. 130–31, 148.
55. For the British figures see Table 1; for the American figures see I.B. Holley, jr., *Buying Aircraft: Materiel Procurement for The Army Air Forces* (Washington, 1964), p. 10.
56. For example, substantial Belgian orders for two Fairey designs were placed on the understanding that the aircraft would be manufactured in Belgium, and 246 Fairey aircraft were eventually built there. Only 171 were built in Britain and exported between the wars. See H.A. Taylor, *Fairey Aircraft since 1915* (London, 1988), pp. 37, 438–42.
57. PRO AIR 2/266, S.23447, D of E to D of C, 21 March 1924; PSO to AMSR, 21 February 1925.
58. PRO AIR 2/266, S.23447, G.P. Bulman to PSO, 11 February 1925.
59. PRO AIR 2/266, S.23447, PSO to AMSR, 21 February 1925.
60. PRO AIR 2/266, S.23447, unsigned memorandum to D of C, 6 November 1924.
61. PRO AIR 2/266, S.23447, PSO to AMSR, 17 April 1925.
62. PRO AIR 2/266, S.23447, D of C to AMSR, 16 July 1925.
63. PRO AIR 2/266, S.23447, PSO to AMSR, 22 January 1926.
64. PRO AIR 2/266, S.23563, AMSR to D of C, 29 May 1924.
65. PRO AIR 2/266, S.23447, AMSR's comments on a memorandum by D of C, 16 July 1925.
66. PRO AIR 2/1213, AMSR to CAS, the Secretary, and S of S, 2 July 1926.
67. Ibid.
68. PRO AIR 2/1213, CAS to the Secretary and S of S, 12 July 1926; CAS to the Secretary and S of S, 30 July 1926.
69. For example, Thomas Sopwith of Hawker, Geoffrey de Havilland, Robert Blackburn, Oswald Short, Roy Dobson and Roy Chadwick of A.V. Roe, Ernest Hives of Rolls-Royce, Roy Fedden of Bristol, Richard Fairey, Sir John Siddeley and Frederick Handley Page.
70. PRO CAB 60/34, memorandum entitled 'Model Plan for the Production of Aircraft in a

National War', sent by the Secretary of the Air Ministry to the Secretary of the Committee of Imperial Defence, 23 August 1927. The plan was prepared by Air Commodore L.E.O. Charlton (PSO).

71. PRO AIR 2/266, S.23447, PSO to AMSR, 17 April 1925; PRO AIR 2/1208, Air Ministry conference to discuss equipment of RAF with all-metal aircraft, 24 November 1925.
72. PRO AIR 2/1213, AMSR to CAS, the Secretary, and S of S, 2 July 1926; PRO CAB 60/34, Secretary of the Air Ministry to Secretary of the Committee of Imperial Defence, 23 August 1927.
73. Ibid.
74. PRO AIR 2/1322, memorandum by D of C, 3 November 1931.
75. PRO AIR 2/1208, reports by D of C, January 1928, February 1929; PRO AIR 2/1322, report by the Aircraft Supply Committee, 2 September 1931, comments by D of C.
76. PRO AIR 2/1322, report by the Aircraft Supply Committee, 2 September 1931, comments by D of C, Secretary and AMSR.
77. Ibid.
78. Ibid.
79. Ibid.
80. PRO AIR 2/1322, comments by Weir, 18 January 1932.
81. PRO AIR 2/1322, summary of AMSR's final comments on recommendations by the Aircraft Supply Committee, AMSR to the Secretary, 12 October 1934. A.V. Roe's deliveries in these years are recorded in PRO AIR 19/524.
82. R.H. Dobson and R.F. Taylor, 'The Jointing of Materials by Welding', *Journal of the Royal Aeronautical Society*, 40 (1936), p. 648.
83. A.J. Jackson, *Avro Aircraft Since 1908* (London, 1965), pp. 287, 292.
84. PRO AIR 2/1790, DAP to AMSO, 11 May 1936.
85. W.J. Reader, *Architect of Air Power: The Life of the First Viscount Weir of Eastwood, 1877–1959* (London, 1968), pp. 193–4.
86. Lloyd, *Years of Endeavour*, p. 150.
87. PRO SUPP 3/44, memorandum by Air Commodore Bigsworth, 15 December 1932; memorandum by Bigsworth, 22 December 1933; 'Extract from 6th annual Report of Supply Committee No. 6' (no date), paper PSO (SB), p. 343.
88. Ibid.
89. PRO AIR 2/2014, report by DAP, 10 January 1938; PRO AVIA 46/31, undated compilation of comparative war potential plans.
90. Weir 19/1, the Secretary to Sir Hardman Lever (chairman of Air Ministry Advisory Committee on Contracts), 6 December 1935.
91. Fearon, 'Aircraft Manufacturing', p. 230; HSA, Armstrong Whitworth board minutes, 12 December 1932.
92. HSA, Blackburn board minutes, 18 July 1935.
93. Letter by C.R. Fairey to *The Aeroplane*, 8 February 1928, quoted in Fearon, 'Airframe Industry', p. 250.
94. Vickers 1392, McLean to Vickers board, 11 February 1935.
95. In a monoblock engine the cylinder blocks were machined from six-cylinder aluminium alloy castings. See Lloyd, *Years of Endeavour*, p. 97.
96. HSA, Blackburn board minutes, 22 December 1925, 4 January 1926.
97. PRO AIR 2/1208, D of C to AMSR, 8 March 1926, January 1930 and 30 January 1931; Vickers 687, J.D. Scott's interviews with McLean, 8 and 9 November 1959. Geodetic design was based on the principle that two geodetic arcs could be drawn to intersect on a streamlined surface in such a way that the torsional loading on each arc would cancel out that of the other.
98. Fearon, 'Formative Years', p. 486; 'Airframe Industry', pp. 239–41.
99. Edgerton, *Aeroplane*, pp. 33–4.
100. C.R. Fairey, 'The Future of Aeroplane Design for the Services', *Royal United Services Institution Journal*, 76 (1931), pp. 574–6.
101. A.J. Robertson, 'Airframe Industry', pp. 649, 656.
102. Vickers 369, McLean to Craven, October 1937; H.J. Tuffen and A.E. Tagg, *The Hawker Hurricane: Design, Development and Production* (Royal Aeronautical Society Historical Group, 1985), p. 9. Tuffen and Tagg worked for Hawker's design and pro-

duction departments during the Second World War.

103. Fearon, 'Airframe Industry', pp. 249–50. Several American airlines and aircraft manufacturing companies were under common ownership.

104. The difference in design practice survived long after the replacement of the biplane by the monoplane in Britain. It was most clearly illustrated by comparisons between British and American aircraft produced during the Second World War but it was already established by the early 1930s. See Vickers 322, memorandum on Vickers-Supermarine technical mission to the USA prepared by Pierson, Westbrook and Shenstone, submitted to Vickers Aviation board on 31 July 1934; Fairey, 'Aeroplane Design', pp. 576, 580; PRO AVIA 10/100, Consolidated report on inspection trip of US aircraft production mission to England, 14 October to 11 November 1942; PRO AVIA 15/2660, report by North American Aviation on weight comparison between Mustang and Spitfire, February 1943.

105. Stressed-skin construction is the use of the outer skin of a structure to carry primary structural loads.

106. PRO AIR 2/1668, meeting between the AMSR and the aircraft industry, 23 November 1934.

107. PRO AIR 2/1668, DTD to AMSR, 15 January 1935; DDTD to DTD, 8 January 1935. The specification G4/31, for a general-purpose aircraft, called for the following capabilities: day and night-bombing, dive bombing, army co-operation, general reconnaissance, aerial photography, casualty evacuation, coastal reconnaissance and torpedo-bombing.

108. Interview with S.D. Davies, 21 April 1990; HSA, Blackburn board minutes, especially 1 November 1922, 18 April 1923, 19 October 1923 (when nine designs were under consideration), 20 September 1926, 16 September 1927, 18 September 1930, 25 September 1934, 4 January 1935.

109. PRO AIR 2/1668, DDTD to DTD, 8 January 1935; DTD to AMSR, 15 November 1935; note for the Air Council by Weir, 21 July 1935.

110. For a detailed breakdown of this timetable see PRO AIR 2/1668. The eight-year development schedule is incorrectly described as an average or even a minimum in the following works: M. Smith, 'Planning and Building the British Bomber Force, 1934–1939', *Business History Review*, 54 (1980), p. 38; Higham, 'Quantity vs. Quality', pp. 449–50. The error first appeared in Postan, Hay and Scott, *Design and Development*, p. 142.

111. PRO CAB 102/51, 'Reciprocating Aero-Engines and Engine Accessories, Production and Programmes', by D A Parry (unpublished official narrative), section 2.

112. PRO AIR 2/1668, H.J. Thomas (Chairman of the SBAC) to the Secretary of the Air Ministry, 28 February 1935; S8 to AMRD, 6 April 1935; extract from Air Council minutes, 7 May 1935.

113. Postan, Hay and Scott, *Design and Development*, pp. 33–4.

114. Vickers 369, McLean to Craven, October 1937. The range with a 1000 lb bomb load was 2550 miles.

115. PRO AVIA 46/116, Official Historian's type biography of the Mosquito.

116. Tuffen and Tagg, *The Hawker Hurricane*, p. 45.

117. J. Morpurgo, *Barnes Wallis* (London, 1972), Chapters 9–11; Vickers 701, J.D. Scott's interview with T.C.L. Westbrook (formerly of Vickers), 15 October 1959; Vickers 780, Scott's interview with Barnes Wallis, 1 February 1960; Weir 19/20, meeting between the Air Ministry and McLean, 6 February 1936; Weir to S of S, 10 September 1936.

118. Fearon, 'Aircraft Manufacturing', pp. 216–25, 230–1; Barnett, *Audit*, p. 130.

2

The Air Ministry and military aircraft production, 1935–39

The mobilisation plans drawn up by the Air Ministry between 1918 and 1934 envisaged a sudden and rapid transition from peacetime conditions to wartime conditions. If this scenario was accurate it seemed certain that there would be a substantial lack of aircraft in the first year of a war, so the planners tried to estimate the extent of this deficiency and to devise means by which it might be overcome. By the onset of rearmament general plans existed for a wartime industry comprising groups of aircraft and engineering firms. One group would be established to produce each of the most important aircraft and aero-engines, and only one aircraft from each of the main operational classes would be selected for group production. An Air Ministry organisation would be created to co-ordinate the output of the different sectors of the aircraft industry.

The planners did not anticipate an intermediate phase, lasting several years, between peacetime and wartime, nor did they expect that a technological revolution would occur in aircraft design. Under these circumstances there were both practical and political objections to the implementation of war production plans, and little scope existed to develop a State production policy. Targets were defined in terms of first-line air force strength rather than industrial capacity. Not until the *Anschluss* in March 1938 was peacetime planning combined with longer-term preparations for total war.

The first expansion programme, scheme A, was introduced in July 1934 but was not scheduled for completion until March 1939. Scheme A was intended to create a Metropolitan Air Force (MAF) of 84 squadrons (1252 first-line aircraft), and involved an increase of 30 per cent in the value of aircraft contracts by comparison with the previous year.[1] Its successor, scheme C, was much more ambitious. Beginning in May 1935, the new plan's objective was to enlarge the

MAF to 1512 first-line aircraft. Approximately 3800 aircraft would
have to be produced by April 1937; only 620 had been delivered to
the Air Ministry in 1934 (see Table 6). These demands were, in fact,
equivalent to the entire capacity of both the military and civil aircraft industries in 1935.

TABLE 6

British Air Rearmament Programmes, 1934–40

Scheme	MAF First-Line Strength	Planned Annual Aircraft Production	Production in Previous Year
A (1934–39)	1,252	n.a.	n.a.
C (1935–37)	1,512	1,900	1,108 (1934)
F (1936–39)	1,736	2,667	1,807 (1935)
L (1938–40)	2,373	6,000	2,218 (1937)

Source: M.M. Postan, *British War Production*, p. 15; J.A. Cross, *Lord
Swinton*, pp. 189–210; 1934 and 1935 production figures from *Census of
Production*, 1935; 1937 production figures from *Statistical Digest of the
War*, p. 152. 1937 production figures relate to Air Ministry orders alone.

Aircraft of modern design were only ordered in small quantities,
however, the contracts for 150 Fairey Battles and 96 Vickers
Wellesleys being typical. In this respect scheme F (March 1936)
involved an advance of far greater significance. Under scheme F,
light bombers like the Hawker Hart were replaced by new monoplanes like the Fairey Battle and Bristol Blenheim. A thousand of
these aircraft were ordered from the designing firms, and even
larger contracts were placed with the so-called 'shadow' industry
which was now created at government expense and managed by
leading automobile manufacturers like Austin and Rootes. In the
longer term, however, the emphasis in bomber procurement was
moving away from the light bomber altogether and towards
medium bombers like the Vickers Wellington and the Handley Page
Hampden, 360 of which were now ordered. Alongside the bombers
came the first monoplane fighters, the Supermarine Spitfire and the
Hawker Hurricane. A contract for 600 Hurricanes was placed with
Hawker in February 1936.[2]

The MAF would now have 1736 first-line aircraft; overseas forces
would be built up to 37 squadrons equipped with 468 first-line aircraft and the Fleet Air Arm would have 26 squadrons with 312 air-

craft. This involved the production of some 8000 aircraft between March 1936 and April 1939 and entailed an increase in average annual output of nearly 50 per cent over the industry's total military and civil output in 1935. A purely numerical comparison understates the effort involved, however. As the aircraft themselves were larger and more complex than those previously ordered they inevitably required many more man-hours to build.

Even scheme F was not intended to provide Britain with anything like a war capability. It differed from scheme C by making adequate provision for reserves and by re-equipping the air force with modern types of aircraft, but its principal aim was, nevertheless, deterrence. The Air Ministry calculated at the end of 1937 that the maximum possible output of the aircraft industry in the first year of a war would be 7080 aircraft – less than half the Air Staff's requirements. An additional five million square feet of manufacturing space would be needed to meet wartime demands; the area actually productive in 1938 would therefore have to be doubled.[3] There was, however, no certainty that the country's industrial infrastructure could support aircraft production on the scale which this implied.[4]

The first proposals for augmenting Britain's expansion plans were made as early as December 1936 and, in October 1937, scheme J was presented to the Cabinet, precipitating five months of wrangling over the size, nature and cost of air rearmament. Following the *Anschluss* the Cabinet accepted a new programme, scheme L. The MAF first line was increased by 37 per cent, but within this total the number of fighters was raised by 45 per cent. This involved the production of 12,000 aircraft in the following two years, an increase in the average annual output of military aircraft of 170 per cent by comparison with the level achieved in 1937. The first-line target was raised still further following the Munich crisis when the decision was taken to create twelve additional fighter squadrons (192 first-line aircraft) and to re-equip Bomber Command with heavy bombers. In the meantime, planning commenced for a War Potential programme for the production of 17,000 aircraft in the first twelve months of a war beginning in October 1939, and for a wartime capacity of 2000 aircraft per month by the end of 1941.[5] Until war-potential planning was sanctioned by the government, industrial capacity had been seen as a means to an end: the creation of a given first-line air force strength. From 1938 onwards the development of industrial capacity became an end in itself.

AIR MINISTRY ORGANISATION AND POLICY, 1935–36

The increasing scale and complexity of aircraft production and the transition from peacetime to wartime planning were reflected in the development of the Air Ministry's production organisation. Until the beginning of scheme C, responsibility for aircraft supply was shared by the Ministry's Directorate of Equipment and the Directorate of Contracts. In May 1935, however, it was proposed that a special staff should be appointed to supervise production orders, and four appointments were subsequently made to the Department of Supply and Organisation for this purpose.[6] Of these, three were given to RAF officers who had worked with the PSOC, while the fourth went to one Owen Clegg, who had appropriate industrial experience.[7] Not until October 1935 did they actually assume their duties, and between them they did not, as yet, constitute a production directorate. They were sometimes called on to report to the Air Council on the progress of aircraft contracts, but they were not in any way responsible for formulating a production policy. This was exclusively the responsibility of the Air Council itself which, from June 1935, held weekly meetings to survey the progress of rearmament.[8]

Meanwhile, a more significant step was taken by the government when, on 18 May 1935, Lord Weir was invited to assume the role of industrial adviser to the Air Ministry. Weir's extensive career in industry had encompassed not only the management of his own companies but also control of aircraft production during the First World War and advisory work for the PSOC. These qualifications alone suggested that he was the best man to assist the Air Ministry with industrial policy but the government also felt that his appointment would inspire public confidence in the rearmament programme. Weir agreed to act in an advisory capacity without undertaking any executive duties. His biographer notes, however, that he was always likely to become 'a good deal more' than a mere adviser.[9]

Weir attended the majority of Air Council progress meetings between 1935 and 1938 but the scope of his authority ultimately depended on his influence over the Secretary of State for Air. Sir Philip Cunliffe-Lister, who chaired the Subcommittee on Air Parity of the government's Defence Requirements Committee in May and succeeded Lord Londonderry at the Air Ministry in June, was a personal friend of Weir and often favoured his advice over that

supplied by senior Air Ministry officials.[10] It was inevitable, therefore, that aircraft production policy between 1935 and 1938 should
have been strongly influenced by Weir's personal philosophy,
which is accurately summarised in his own notes. He wrote:

I was definitely averse to doing anything which would turn industry upside
down by creating a war spirit and practice, but I felt that we must quietly
but very rapidly find an effective British compromise solution as opposed
to merely copying the dictator system.[11]

This outlook did not prevent a state production policy, which
went well beyond merely placing orders, from emerging during
1935. It did, however, ensure that the government's role was
limited to the provision of incentives to the industry to expand its
capacity. The Air Ministry believed that the existing aircraft
industry was capable of fulfilling the requirements of scheme C, but
there was some anxiety about the division of the programme
between the two financial years, 1935–36 and 1936–37. Contracts
could not normally have been placed for each year until the annual
defence estimates had been approved. However, scheme C, unlike
its predecessor, presented an opportunity to organise uninterrupted
production programmes over two years, with consequent advantages in terms of both cost and efficiency. The Subcommittee on Air
Parity therefore urged a relaxation of Treasury and parliamentary
controls over annual expenditure by recommending

that in the special circumstances the difficulties which normally arise each
year in getting contracts placed in sufficient time for them to be completed
must be overcome by the authority which the Air Ministry require and by
the co-operation of all departments concerned.[12]

The extended production programme thus became the corner-
stone of the government's aircraft production policy.[13] It was a
limited policy, but a policy with important industrial implications,
as Air Ministry research in the 1920s and early 1930s had suggested.
In particular, the Ministry hoped that the longer-term contracts
would encourage the industry to jig and tool much more extensively, which they did. The same reasoning lay behind the allocation
of three-year programmes for scheme F in the following year. It
took time for the large-scale tooling envisaged by the Air Ministry
to be completed but the higher degree of mechanisation ultimately
allowed production to be increased very rapidly during 1938 and
1939.[14]

There were two corollaries of the Air Ministry's policy of

working through the professional aircraft industry. First, the firms would have to expand their own factories. Second, provision would have to be made for the increased working capital which the industry would require. Although the government viewed the expansion of the aircraft industry as essential preparation for the event of war, it did not, as yet, consider that the state should provide buildings, plant, or additional finance, and during the course of 1935 a number of firms took the initiative by floating new issues, enlarging their factories, or purchasing new ones.

The industrialists were nevertheless haunted by the spectre of over-capitalisation and redundant capacity.[15] Following representations by the SBAC in October 1935, the Air Ministry gave assurances that compensation would be paid for capital extensions which became redundant when the expansion programme came to an end, and in the same month concessions were also made to the industry on the question of working capital. The Ministry agreed to make monthly payments of 80 per cent of the money spent by contractors on the purchase of materials, and to make so-called 'progress payments' for materials amounting to 90 per cent of their estimated total cost.[16] The compensation agreement (soon known as the Capital Clause) and the progress payment scheme were later extended to cover subsequent rearmament programmes.[17]

These contractual concessions marked the limit of state production policy, and the Air Ministry's supply functions were otherwise restricted to monitoring the progress of individual contracts and bringing pressure to bear on those contractors who fell behind with deliveries. In the case of one firm, Westland, the Air Ministry briefly intervened in October 1935 when it appeared that the company was on the verge of collapse. In this instance the ultimate sanction of threatening to withhold orders was employed, and the Ministry undertook its own investigation of Westland's finances.[18] But this was the only occasion on which any such action was warranted during scheme C.

AIR MINISTRY ORGANISATION, 1936–38

Following the approval of scheme F in 1936, a new directorate specifically charged with the supervision of aircraft production was established in the Department of Supply and Organisation. It was named the Directorate of Aeronautical Production (DAP). At the head of the DAP was Colonel Henry Disney. Disney was educated at Marlborough and Cambridge and had served in the Royal Flying Corps during the Great War. Since 1918 he had held directorships at several telegraph, telephone, and cable companies. His combination of service and business experience seemed to Newall to be an advantage, and Weir also recommended that Disney should be appointed. 'He appears', Weir wrote, 'to have been successful in his business career, especially in reorganising and clearing up inefficiencies. He is not in any degree an engineering production expert, but he knows the general factors involved.'[19] In April 1936 Disney had a staff of seven, drawn mainly from the RAF. As the Directorate expanded over the next year the proportion of air force officers declined, and by July 1938 the majority of the 27 staff in the DAP had been recruited from outside the air force. Most, presumably, had business or Civil Service backgrounds but only four are listed as members of professional engineering bodies.

Circumstances conspired to limit the power of the DAP. Most of the important selection decisions had been taken before Disney had time to make his presence felt, but he was not, in any case, given sufficient authority to influence the selection process.[20] He did not have any powers to supervise factory planning, for until 1938 the overwhelming preponderance of capital investment in the aircraft industry was privately financed; nor was the DAP empowered to initiate industrial planning for war, as scheme F had to be completed within a relatively short space of time and did not involve the creation of a long-term war potential. In an attempt to meet the April 1939 deadline, Lord Swinton (as Cunliffe-Lister became in November 1935) agreed to the 'shadow' aero-engine industry being organised in a manner to which the DAP was irrevocably opposed.[21] And in 1937 Newall's Private Secretary thwarted the directorate's recommendations for creating war potential capacity for Rolls-Royce aero-engines, arguing that the company's existing factory was adequate for scheme F and that, in wartime, spare capacity could be transferred to Rolls-Royce from the automobile industry.[22]

The strongest constraint on the DAP's authority, however, was imposed not by his superiors in the Air Ministry, nor by government policy, but by the technical revolution then occurring in aircraft design. Guiding a new engineering project through the design process and then into production was (and is) a very difficult industrial operation. It was particularly difficult for the aircraft industry in the mid-1930s because of the transition from the biplane to the monoplane. The industry was developing a completely new technology. As the government did not design and build its own aircraft, the problems involved could only be solved at the industrial level, and this took time.

The industry's progress will be discussed in the following chapter. Suffice it to say that in the later months of 1936 and in 1937 the delivery schedules for the new types which the RAF was most anxious to obtain began to slip back. In such circumstances it proved impossible for the DAP to undertake the duties which might otherwise have been expected of a state production directorate.

For example, the DAP was unable to develop reliable production programmes. Programmes could not be calculated scientifically on the basis of labour, floor space or machinery because this capacity was impossible to employ efficiently until new types of aircraft had been standardised for production. Nor was there any scientific basis for predicting the duration of the standardisation process itself. In the words of one senior Air Ministry official:

there was no yardstick of previous experience by which production could be measured, and neither the Air Ministry nor the firms themselves had any idea how long the preparatory stages to quantity production took. The optimistic figures quoted by the firms were pure guesswork and the Air Ministry knew no better.[23]

The DAP did monitor the weekly output of the airframe and aero-engine industries, and its estimates of the ultimate production levels achieved during scheme F proved reasonably accurate.[24] A statistician was appointed during 1937, and when production arrangements were finalised by the firms Disney's staff could base their calculations on the number of assembly jigs and on the anticipated output per jig.[25]

However, as far as the commencement of production was concerned, the directorate's predictions proved hopelessly optimistic. Moreover, the concept of programming could only be developed on a very restricted scale because of the time limit which had been imposed on scheme F. The objective was a particular first-line air force strength. The scheme was not intended to create the industrial capacity necessary for a given monthly or yearly output. It was therefore less important to balance the monthly production of airframes, engines, and equipment than to ensure that sufficient cumulative deliveries had been made by April 1939. At the time of the Munich crisis Rolls-Royce had delivered 1700 Merlin engines of which just 600 had been installed into aircraft.[26]

If the DAP had only minimal success in drawing up programmes there was also very little that it could do to accelerate output. The key bottleneck in the production process was the industry's design capacity, and the distribution of work among the various technical teams lay beyond the directorate's jurisdiction. At the beginning of rearmament, the firms were literally swamped with experimental projects. By October 1936 no fewer than 55 different experimental designs were in progress and six more were under consideration.[27] The acute pressure under which the industry was working manifested itself in design errors which were not discovered until production started. The resulting disorganisation was then seized on by the DAP as evidence of inefficiency.[28]

The Air Ministry did not immediately recognise that it was itself contributing to the most critical of the industry's difficulties. Not until December 1937 did the AMRD (who had no formal responsibility for production) recommend giving more consideration to the firms' technical capacity when apportioning experimental work.[29]

The production shortfalls of 1936 and 1937 did lead the Air Ministry to insist on a number of managerial changes which were probably long overdue.[30] The DAP had no powers to compel firms to implement particular reforms if they were strongly opposed to doing so, however. If contractors rejected the DAP's advice and failed to find alternative solutions to their problems, Disney alerted the Air Council and the AMSO then intervened directly with the senior executives of the company concerned. If there was still no improvement, the matter was taken up by Weir or Swinton.

Air Ministry intervention, however, did not always help matters. Some industrialists thought that the formation of the DAP was a precursor to nationalisation, which was then under way in France, and the directorate encountered resentment and hostility from all quarters.[31] The Secretary of State also provoked the industry's antagonism. As one senior official later recalled, after meetings with Swinton the representatives of the firms invariably 'went away feeling aggrieved and that they had been badly treated by the Air Ministry.'[32] In fact Swinton recognised that state interference could prove counterproductive and preferred the Air Ministry to work in a purely advisory capacity if at all possible.[33] Weir agreed: 'I am confident', he wrote, 'that it is wiser to let the different units of the industry continue to work hard and actively at solving their works management and production problems themselves.'[34]

In the later months of 1937 the slow progress and high cost of rearmament began to cause political controversy and the Air Ministry was forced to consider extending the DAP's authority. The Ministry's Secretariat, which was responsible for financial matters, argued that the DAP should not simply report on production problems but should take active steps to resolve them. It is unclear whether anyone in the directorate was actually qualified to do this but, in any case, Swinton remained cautious. He did eventually agree to strengthen Disney's staff by appointing resident progress officers (so-called 'overseers') to several important aircraft factories,[35] but otherwise the Secretary of State professed himself reasonably satisfied with the management of the industry. He maintained that the DAP had been established to 'chase' production and that it could not insist on the adoption of particular production methods.[36] Again he was supported by Weir. The aircraft industry was, Weir wrote, a young industry 'developing its scientific possibilities . . . which demands encouragement of the imagination and enterprise of the individual'.[37]

Between 1936 and 1938, then, strict limitations were imposed on the authority of the Directorate of Aeronautical Production, partly by its position within the Air Ministry's organisational hierarchy, partly by government policy, but chiefly by the revolution in aircraft design and its various industrial repercussions. There was minimal scope for developing a state production policy until the design and development of new aircraft had been completed; the only alternative was to leave the aircraft industry to its own devices. Swinton and Weir were correct to advise restraint.

During 1938 the position changed decisively. Following the *Anschluss*, scheme L was launched, and war-potential planning commenced. The new designs ordered in 1936 began to emerge in large quantities. For the first time since the beginning of rearmament, production *per se* became an objective. A complete review was therefore undertaken of the Air Ministry's supply organisation. To improve co-ordination between the Ministry's design and production activities the departments of research and development and supply were reunited and placed under Dowding's successor as AMRD, Air Marshal Sir Wilfrid Freeman, who became the Air Member for Development and Production (AMDP) in June.[38] Finally, responsibility for the implementation of scheme L was given to an Air Council subcommittee which soon became known as the Air Council Subcommittee on Supply.[39]

Following the resignation of Swinton and Weir in May 1938 and the appointment of Sir Kingsley Wood as Secretary of State for Air, the Supply Committee virtually assumed complete responsibility for aircraft production. 'The Air Supply Board spent [the] money,' Freeman later recalled, 'Kingsley Wood left us alone.'[40] The Supply Committee was primarily established to apportion contracts under scheme L. As the government had now agreed to finance extensions to the aircraft factories, however, the allocation of such expenditure also became a function of the committee. Indeed, this became its principal role after July 1938 when it was decided that capacity should be planned and constructed for the War Potential programme.

To enable the Air Ministry to reach rapid decisions on both contracts and factory extensions, a permanent seat on the Supply Committee was allocated to the Treasury.[41] Another member was the Chairman of the SBAC, Sir Charles Bruce-Gardner. The firms supplied Bruce-Gardner with estimates of the maximum number of aircraft which they could deliver by April 1940, and these proposals formed the basis of subsequent discussions between the Air Ministry and the aircraft industry, whose senior executives regularly attended the Supply Committee during the following months.[42] The consultative process proved far more effective than the series of *ad hoc* meetings which had taken place between the two sides during the previous three years, and compelled both to take a much more realistic view of the length of time required to put new types of aircraft into production.

For the following two years, Sir Wilfrid Freeman presided over aircraft production policy. Freeman was educated at Rugby and Sandhurst and, after joining the Manchester Regiment, he was seconded to the RFC in 1913. After the First World War he joined the RAF as a Wing Commander and, having progressed through the senior ranks, he was appointed to the Air Council in 1936 as Member for Research and Development.[43] His personal talents were always more in evidence in aircraft design and development than in production, but at the end of 1937 he had been involved in a series of discussions with the industry on accelerating the introduction of new types of aircraft and, at about the same time, he was making his own recommendations to the Air Council on production matters even though they were, strictly speaking, beyond his jurisdiction.[44]

With aircraft procurement occupying such an important position in the rearmament programme it made practical sense to place the AMSO's production responsibilities under Freeman, as opposed to transferring aircraft research and development to the AMSO. The new Department of Development and Production became, in the words of the official historian, 'the embryo of the Ministry of Aircraft Production of future years'.[45]

Freeman had no formal industrial training but this was not necessarily a disadvantage. He was charged not with formulating central production plans but with co-ordinating production possibilities with the demands of the user, the air force. The responsibilities of the new department were, in any case, so numerous that it would have been impossible for him to have supervised either its development or production activities in detail. Day-to-day control therefore passed to two new directorates. Air Vice-Marshal Tedder became the Director General of Research and Development (DGRD), while one of Britain's leading planning engineers, Ernest Lemon, accepted the appointment of Director General of Production (DGP).

The son of an agricultural labourer, Lemon gained his engineering qualifications by combining an apprenticeship with part-time study at Glasgow Technical College and at Heriot-Watt College, Edinburgh. He made his career managing the production of railway carriages but subsequently became involved in overall railway management for LMS, becoming Operating and Commercial Vice-President in 1932. He was approached by the Air Ministry following the recommendation of a panel of industrial advisers appointed by Kingsley Wood in June 1938.[46]

Under Lemon, the Air Ministry's production organisation was transformed. The DAP's functions were restricted to airframes alone, other responsibilities being divided between several new directorates with jurisdiction over aero-engines, equipment, subcontracting, materials, factory construction, statistics and planning, and war production (see Figure 1). The work of the different directorates was then co-ordinated through weekly meetings chaired by the Director General and, in 1939, an aircraft production programme balancing the monthly output of airframes with requirements for ancillary items was drawn up with the assistance of planning experts from one of the principal aircraft equipment manufacturers, Joseph Lucas Ltd. An Air Ministry organisation was subsequently created to extend the aircraft production programme into 800 main items of aircraft equipment.[47]

How successful was the new administration? The remainder of this chapter examines some of the key areas of production policy –

the creation of group production schemes, the expansion of the 'shadow' industry, labour supply, and the provision of aircraft equipment and aluminium – in an attempt to answer this question.

FIGURE 1

The Air Ministry's Production Organisation, 1934–40

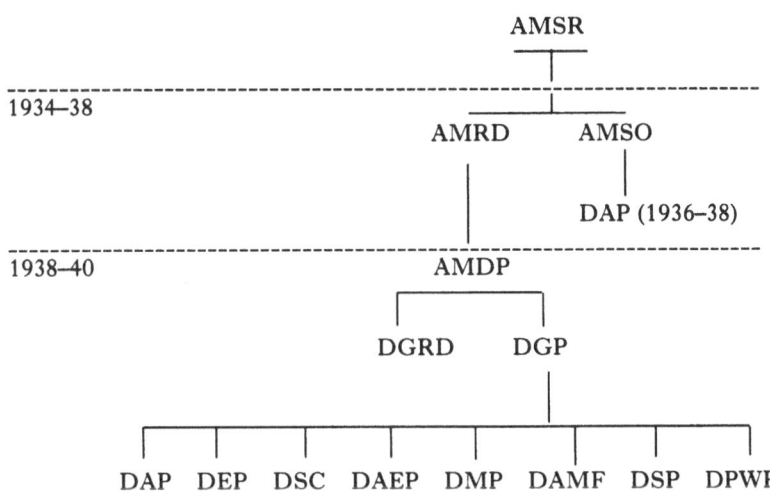

Key

AMSR:	Air Member for Supply and Research
AMSO:	Air Member for Supply and Organisation
AMRD:	Air Member for Research and Development
DAP:	Director of Aeronautical Production
DGP:	Director General of Production
DGRD:	Director General of Research and Development
DEP:	Director General of Engine Production
DSC:	Director of Subcontracting
DAEP:	Director of Aircraft Equipment Production
DMP:	Director of Materials Production
DAMF:	Director of Air Ministry Factories
DSP:	Director of Statistics and Planning
DPWP:	Director of War Production Planning

PRODUCTION POLICY

The group schemes

As already noted, the plans devised by the Air Ministry prior to rearmament envisaged that, in wartime, aircraft and engineering companies would be organised into groups. Each group would manufacture a particular type of aircraft and only one type would be ordered from each operational class. Certain officials assumed that this approach would be adopted at the beginning of scheme F, but Weir took a different view, arguing that group production was 'conceived on a basis of war preparation and it would not appear to help us very much in our 3/5 year programme'.[48]

There were numerous difficulties. As the group system had been formulated with war production in mind, it was always assumed that the government would be in possession of compulsory powers. In fact no such powers were taken, and the expansion of the aircraft industry was initially financed by private capital. This gave the Air Ministry much less influence over the way production was organised than might have been the case if controls had been introduced or if the state had provided fixed capital for the firms. Even if the state had taken more powers, the essential prerequisite to group production – standardisation of design – was lacking. As Bullock wrote in 1936, 'to arrange for the sub-contracting of complete aircraft of new designs is apt to be a lengthy business'.[49] The Air Ministry was, of course, hoping to obtain the new designs as quickly as possible.

The April 1939 deadline for scheme F created other planning problems. As individual firms often lacked the capacity to fulfil all the RAF's requirements, the Air Ministry was forced to order different types to perform the same military functions. For example, the Air Staff required 900 single-seater fighter aircraft for scheme F. Of the two types under consideration, only Hawker's design had flown by February 1936, yet production could not commence before the middle of 1937 and it was not thought that Hawker could deliver more than 600 aircraft by April 1939. So the Air Ministry chose a completely different model, the Supermarine Spitfire, to make up the balance of 300 and Britain went to war with production divided between two fighters when, ideally, resources should have been concentrated on one.[50] Concentration would, however, have eliminated the better aircraft, the Spitfire.

Another practical obstacle to the goal of concentration was the lack of standardisation in construction methods. For example, the Air Ministry had ordered two prototypes to the medium bomber specification B9/32, 360 of which were required for scheme F. Ideally, from the production standpoint, the best design (submitted by Vickers) would have been selected and built by firms with available capacity. However, the Vickers B9/32 (the Wellington) incorporated geodetic construction which required specialised rolling and milling plant, only available at Vickers' Weybridge factory. Vickers already had production contracts for the Wellesley and did not have the capacity to produce more than 180 B9/32s by April 1939. Moreover, geodetic construction was thought to be so revolutionary that production would have to be 'proved' at Vickers before any subcontracting was undertaken.[51]

The Air Ministry's solution was to order the other B9/32 (the Handley Page Hampden) as well, and as a result resources were divided between the two aircraft until the end of 1941.[52] By 1937, the only firms not engaged in the production of their own designs were building obsolete Hawker types ordered for scheme C, and although Boulton Paul, De Havilland and Westland did undertake some subcontract work, not one of the new monoplanes was produced by a non-design firm from the Air Ministry's list of main contractors until July 1938. The highest degree of concentration was achieved on the Blenheim, which was built by Bristol, Rootes and A.V. Roe. A Bolton firm, Dobson and Barlow, assembled wings for all three factories.[53]

Circumstances changed during 1938. The state began to finance fixed investment in the aircraft industry, thereby gaining more control over the way resources were used; designs ordered in 1936 were standardised for production, and a number of firms had by this time gained some experience in producing the new monoplanes. Following the Munich crisis, a series of group committees was therefore established to co-ordinate the production of jigs and tools, material procurement and subcontracting arrangements for several new types of aircraft.[54] By then the Air Ministry was sufficiently sure of the Wellington to concentrate medium bomber production on this one model, though other models remained in production pending the emergence of the new heavy bombers which were then being developed.[55] Plans were also sanctioned which resulted in the bulk of wartime aero-engine production being focused on the Rolls-Royce Merlin and the Bristol Hercules.

It nevertheless remained extremely difficult to concentrate airframe production in the way the planners had envisaged before rearmament. In selecting a standard heavy bomber, the Air Ministry found itself facing many of the same dilemmas that had complicated the selection process in 1936, for not one of the three 'heavies' (the Short Stirling, the Handley Page Halifax and A.V. Roe's Manchester) had flown at the time the group schemes were adopted. The ministry was eager to advance production arrangements and to concentrate resources on one preferred model, but was also fearful of selecting an unsatisfactory model. The design firms each wanted to continue with their own projects.[56]

The result was that groups were organised for all three aircraft with a view to concentrating resources on the best one at a later stage.[57] This was an expensive decision from the production standpoint but it ultimately ensured that Britain developed a first-class heavy bomber. Germany, on the other hand, gambled everything on one such design (the Heinkel He. 177) which failed, leaving the Third Reich without any heavy bombing capability at all.[58] In Britain's case an early decision would almost certainly have eliminated the Manchester, which in 1939 seemed the least promising of the three aircraft but which, having been developed into the Lancaster, proved to be not only the best design but the easiest to produce.[59]

It was equally difficult to concentrate production on a single type of fighter. A group scheme was initially planned for a new Hawker fighter which was intended to replace the Spitfire and the Hurricane in 1941, yet the new aircraft, the Tornado/Typhoon, relied on the successful development of one of two novel and highly complex aero-engines, whereas it became clear at the end of 1939 that a Spitfire with a more conventional power plant could equal the performance of the Hawker design.[60] Consequently it was decided to continue with Spitfire development.[61]

In the meantime the new expansion plans launched in 1938 provided for an increase in first-line fighter strength of 90 per cent above the scheme F objective. The new targets left the Air Ministry with no choice but to commit far greater resources to fighters which were already in production. A Spitfire shadow factory was sanctioned, Supermarine's capacity was enlarged, and a Hurricane group was set up comprising Hawker, Gloster Aircraft, two Scottish engineering companies, and the Canadian Car and Foundry Company in Montreal.[62] So, by the outbreak of war, a substantial industrial commitment had been made to at least three different fighters. At the time it was expected that capacity would later be transferred from the Spitfire and the Hurricane to more modern designs. In the event the Tornado/Typhoon was consistently outclassed by later developments of the Spitfire, which therefore remained at the centre of Britain's fighter production plans for the duration of hostilities.[63]

In its reluctance to standardise airframe production the Air Ministry was ultimately placing quality before quantity. The dangers inherent in standardisation were soon illustrated by a series of costly failures which left the Fleet Air Arm and Coastal Command desperately short of modern aircraft.[64] However, just as the success of two similar designs tended to frustrate plans for greater concentration, so too did the total failure of individual projects advance progress towards that end, with consequent advantages in terms of efficiency. Thus Blackburn, Westland, Saunders-Roe and Boulton Paul spent much of the war producing aircraft designed by other firms. Boulton Paul joined Fairey in the production of the Barracuda torpedo-bomber, and Blackburn also built the Barracuda, the Short Sunderland flying boat, and the Fairey Swordfish torpedo-bomber. Westland produced the Spitfire and Seafire, and Saunders-Roe built Supermarine seaplanes.[65]

The group schemes were substantially revised between 1939 and 1941. The composition of the groups changed, as did some of the aircraft which they were intended to produce. Nevertheless, the five principal groups established by the Air Ministry for the Wellington, Halifax and Manchester bombers and for the Merlin and Hercules engines ultimately absorbed more than half the resources of the wartime aircraft industry.

The shadow schemes

The introduction of scheme F in 1936 led the government to abandon its policy of working only through the established aircraft industry. A new 'shadow' industry was created. It was always seen as a useful war potential, but the initial decision to seek the assistance of the motor industry was dictated entirely by the time limit which the government had set for the peacetime rearmament programme. This affected both the quantity and quality of the aircraft with which the RAF was equipped on the outbreak of war in 1939.

When scheme F was drawn up at the end of 1935, the Air Ministry envisaged spending £29 million on aircraft in 1937 and 1938 and it was thought that the enlargement of the industry that had been undertaken for scheme C would be sufficient to secure the numbers of aircraft and engines involved. Shortly before scheme F was presented to the Cabinet, however, the Chief of the Air Staff pointed out that these plans would leave a substantial amount of expenditure to be carried into 1939 and subsequent years, whereas his instructions were to complete the RAF's first-line preparations by the end of the financial year 1938 (31 March 1939). In consequence, no less than £12 million was pushed back into the two previous years. This was done 'quite arbitrarily and without working out any of its implications – a species of crazy finance which was unfortunate, but inevitable in the circumstances'.[66]

The revised financial plans had then to be translated back into aircraft. The Air Ministry had hoped, somewhat optimistically, for an output of approximately 3000 aircraft in 1937 but the new cash figures implied a rate closer to 4000. The Secretary, Sir Christopher Bullock, considered it 'most improbable that an output of 4000 aircraft per annum can be attained without going outside the existing industry.' In other words, the government was about to allocate a sum of money to aircraft production which could not possibly be spent between 1936 and 1939 because the necessary industrial capacity did not exist. Bullock therefore proposed that the additional aircraft be obtained from those automobile firms which the Committee of Imperial Defence had allotted to the Air Ministry as war potential, and this recommendation was subsequently accepted by the Secretary of State.[67]

There were both advantages and disadvantages involved in the implementation of a shadow scheme at this time. The Air Ministry and its contractors gained invaluable experience of the problems which accompanied shadow production but the lessons were only learned at considerable cost. The shadow industry had to be planned to produce established designs so that drawings and process sheets were available from the outset[68] but the new factories took time to build, equip and staff. As aeronautical technology was advancing rapidly there was thus a danger that the type of aircraft ordered would become obsolete by the time production started.[69]

This was not a problem unique to Britain. During the Second World War the German company, Adam Opel, took so long to prepare for aircraft production that tooling had only been completed for an obsolescent type shortly before the firm was directed to tool up for a newer model.[70] In Britain, the motor industry was initially required to assist with the creation of the 70-squadron bomber force which was the principal objective of scheme F. Yet the only modern bomber designs near completion in 1936 were the Battle and the Blenheim. By the time shadow production started in 1938 the Battle was obsolete and the Blenheim had been superseded by more modern designs like the Vickers Wellington.[71]

The organisation of shadow production proved no less perplexing for the Air Ministry. The most rapid and economical method of producing the 4000 Bristol Mercury engines required for the Blenheim (and several other aircraft) by April 1939 – and the method favoured by Bristol and the leading motor firms – was for each of the seven participating companies to concentrate on one of seven sections of the engine. This would minimise the technical and supervisory burden on Bristol and avoid the duplication of jigs, tools and gauges at a time when Britain's tool-making capacity was being stretched to the limit. On the other hand, the Air Ministry wished to educate the motor industry for the production of complete aero-engines in wartime, and therefore favoured a scheme whereby each firm would have made one of two sections of the engine.[72]

Disney believed the seven-unit scheme (which was eventually reduced to five units after the departure of Wolseley and Singer) would be cumbersome, expensive in terms of management fees, and vulnerable to air attack.[73] The time factor nevertheless persuaded Swinton and Weir to defer to the industrialists. Where the airframe factories were concerned, however, the opposite view prevailed. The new works were entirely self-contained, planned on an elaborate scale and instructed to jig and tool for unlimited numbers of aircraft.[74] This was more than justified at Rootes' Blenheim factory, but tooling expenditure at Austin's was disproportionately high in relation to the total cost of the 1000 Battles produced there.[75]

The Air Ministry's early experiences of shadow production were therefore less than satisfactory. The Ministry did not, however, lose faith in the shadow concept, and the industry was enlarged following the commencement of scheme L. No longer was shadow capacity merely required to help fulfil the peacetime aircraft programme; the new factories were planned as war potential in the expectation of producing large numbers of standardised aircraft. The Nuffield organisation agreed to manage a Spitfire shadow factory at Castle Bromwich capable of producing 240 aircraft per month; the armaments manufacturers, Vickers-Armstrong, assumed responsibility for managing two new assembly plants for the production of the Wellington;[76] the two airframe shadow factories, Austin and Rootes, were enlarged and were joined by English Electric and Metropolitan Vickers in new heavy bomber production groups;[77] a second shadow aero-engine scheme, involving Daimler, Standard, Rootes and Rover, was set up to produce Bristol's Hercules engine;[78] and in 1939 work commenced on a Merlin engine shadow factory in Manchester under the management of the Ford Motor Company.[79] Neither the heavy bombers nor the Hercules engine had been fully developed when these later schemes were implemented but the existing professional and shadow industry was rapidly gaining experience in the manufacture of new types of aircraft and engines. There were still risks involved in farming out undeveloped models but these had now to be accepted to reduce the danger of aircraft becoming obsolete before production began.

New factories were also constructed at government expense for management by the professional aircraft industry. This went against the grain of Air Ministry thinking in the pre-rearmament years but was, nevertheless, essential because of the sheer scope of the acceleration in production required after 1938. Moreover, the experience of the first shadow schemes suggested that quality and quantity could be more effectively co-ordinated if close links were maintained between design and production capacity.[80] Hence wartime output of such important products as the Lancaster bomber and the Merlin aero-engine came mainly from factories managed by the design firms, A.V. Roe and Rolls-Royce, and built at Air Ministry expense between 1938 and 1940. These firms were, however, assisted by very large numbers of subcontractors from outside the aircraft industry.

The new schemes proved far more successful than their predecessors in creating mass-production capacity for up-to-date first-line aircraft and aero-engines. This was partly because of the Air Ministry's willingness to gamble on undeveloped models and partly because the majority of professional firms proved capable of shouldering much greater managerial responsibilities than the Ministry initially expected. But the principal reason was simply that the rate of obsolescence, which was particularly high in the mid-1930s, tended to slow during the early years of the Second World

War, so that modified versions of designs which first appeared in 1938 and 1939 retained their operational utility until 1945.[81]

Labour supply

The Air Ministry began to interest itself in the issue of labour supply after scheme C was sanctioned by the government in 1935. Given the absence of state controls, it seemed certain that the supply of skilled labour would not keep pace with the demand created by the defence programme. Joint consultations between the armed service departments and the Ministry of Labour resulted in the creation of additional capacity for engineering trainees at the government's training centres,[82] but there were some who advocated a more active labour supply policy. The Air Ministry planners had always expected that, in wartime, production would be taken to available labour by subcontracting, and it was the Prime Minister himself who suggested to the DRC 'making aircraft in parts and then assembling them' – in effect, subcontracting production.[83] In July 1935, Sir Alexander Ramsay, a director of the Engineering Employers' Federation (EEF), made a similar suggestion to the Air Council. There was, he argued, 'a large part of aircraft production which could be done by the general engineering industry and which would have to be done by that industry in war.' The Ministry should immediately extend its existing system of central purchasing beyond so-called 'embodiment loan' equipment to airframe components, jigs and tools.[84]

As well as solving labour supply problems, such a scheme would have had the advantage of creating more assembly capacity in the aircraft industry, which, in turn, would have allowed the Air Ministry to order fewer designs in larger quantities. But there were also several objections to this approach. Firstly, while Ramsay argued that the general engineering industry should be educated for war production, the Air Ministry believed that the needs of war production would best be served by the expansion of the professional aircraft firms. The Ministry had already taken the decision to work through its approved contractors, leaving them to subcontract as necessary. It was felt that schemes C and F provided a valuable opportunity for the industry to gain experience in quantity production methods.[85] In any case, the Air Ministry had always been reluctant to subcontract airframe manufacture until production designs had been standardised.[86] It was very much simpler for the firms to 'debug' new designs in their own factories than for modifications to be incorporated by a network of subcontractors, all of whom had to be supervised. Ramsay, of course, knew very little about the technical changes which were then taking place in aircraft design.

So there were cogent arguments against subcontracting during the early stages of rearmament. The Secretary of State for Air expressed the view that it represented 'a much more complicated question which would have to be dealt with over time'.[87] Weir acknowledged that 'in war the system of centralised purchase is essential', but he felt that 'peace conditions may well warrant an entirely different policy, especially in regard to supplies of a highly technical character'. Bullock also expressed the view that the EEF's suggestions should not be pursued.[88]

The issue was not pressing for labour supply was generally adequate during 1936 and 1937, but it was clear that shortages would occur if rearmament was accelerated. As a greater degree of design standardisation was attained, the Air Ministry's attitude towards subcontracting therefore began to change. In December 1937, Freeman suggested creating 'a new satellite sub-contracting organisation' to help fulfil the next expansion plan.[89] Disney agreed. 'Unless something drastic is done in connection with this question of sub-contracting', he warned, 'we shall not attain anything like the aircraft we hoped for under scheme L.'[90]

Weir remained uncertain. Following the *Anschluss* he recommended the introduction of a much more vigorous production policy but he still opposed subcontracting. Swinton also preferred to concentrate production on the professional and shadow aircraft industry. Subcontracting, he told the Cabinet, would cause considerable difficulties because 'contracts . . . were all placed, and the various works laid out, jigged and tooled to deal with the whole contract'.[91] The alternative was to arrange double shifts at all aircraft factories but this immediately raised the question of recruiting additional labour.[92] On 28 March and 5 April, discussions were held with the EEF about the possibility of transferring workers to factories engaged in rearmament work. The government encountered entrenched opposition and the question was shelved pending consultations on labour dilution between the EEF and the engineering unions.[93] No further progress was made before Swinton and Weir left the Air Ministry on 16 May.

Scheme L was based on hurried estimates made by the aircraft firms of the practicability of extending their plant and machinery to accommodate more labour. The assumption was that adequate supplies of labour would be available. However, their calculations were believed by the Supply Committee to be somewhat optimistic, and in several cases the estimates were reduced before orders were placed. Yet Freeman was, even then, doubtful of the firms' capacity to fulfil the new production targets. 'I feel that a more detailed investigation . . . is essential', he wrote, 'to ascertain the weak spots in each programme so that prior steps can be taken to avoid a breakdown.' He went on to suggest that statistical information should be compiled showing the existing and future requirements of each firm in relation to floor space, labour, machine tools, equipment and material.[94]

The investigation became Lemon's primary objective between June and August 1938. He was assisted by a Canadian production engineer, Lewis Ord, and later by a British management consultant, T.S. Smith, who became Director of the Air Ministry's new Statistics and Planning section. Although not as comprehensive as Freeman initially intended, Lemon's report was sufficiently thorough to demonstrate that the floor space and machining capacity at most airframe firms was adequate for the peacetime programme and to focus attention on the problem of labour supply.[95]

The consultations between the EEF and the engineering unions in May failed to produce any agreement over dilution. This did not prevent some de-skilling and upgrading by the aircraft firms and, as the volume of output increased, the ratio of semi-skilled labour to skilled labour tended to rise. Moreover, recession in the engineering industry had helped to improve the supply of skilled labour somewhat.[96] Yet by August 1938, the size of the work-force was only increasing by six per cent per month and Lemon considered this totally inadequate. With the EEF opposing the introduction of labour transfers, and the unions resisting dilution, the only alternative was to distribute aircraft work to the general engineering industry.[97] Airframe firms were therefore encouraged to subcontract 35 per cent of productive man-hours and to set up subcontract departments to supervise outside firms.[98] At the same time a directorate of subcontracting was created within Lemon's department to establish the suitability of potential subcontractors,[99] and nearly all new aircraft factories sanctioned between 1938 and 1940 were planned on the basis that a large proportion of work would be subcontracted.[100]

Not that Lemon's staff was unanimously agreed that subcontracting provided the answer to the labour problem. Ord believed strongly that the productivity of the existing labour force could be increased: 'Subcontracting . . . would be bound to lower the efficiency and raise average manufacturing costs. This would make a bad situation worse.' He argued that, while subcontracting was being extended, the Ministry should therefore seek to establish the causes of low productivity at some of the firms.[101] During 1939 two detailed studies of production costs were in fact undertaken.[102]

But in 1938 the production figures did not provide a very reliable basis for measuring labour productivity. The two shadow airframe factories, both Vickers factories, Handley Page, Short Harland, and A.V. Roe's Blenheim factory were not in full production, and in 1939 the industry proved capable of a much higher output than it achieved in 1938 without any pressure from the Air Ministry to increase efficiency. Lemon's investigation, made in the summer of 1938, exaggerated the likely shortfall in labour supplies. 'I rather stressed the labour point', he wrote to Frederick Handley Page, 'because I did not see at the time it was possible for people to so reorganise their factories that the labour force would not require to go up proportionately to the number of man-hours required per machine.'[103] Nevertheless, the DGP's plans for subcontracting progressed rapidly during the following 12 months. By July 1939, more than 1200 firms were engaged in airframe subcontracts, accounting for 15 per cent of airframe labour.[104]

Aircraft equipment and aluminium

Between 1935 and 1938, airframe production represented the single most troublesome constraint on the entire aircraft programme and it was inevitable that the Air Ministry's efforts to accelerate production should have been focused on the airframe sector. Yet with the development of mechanically operated gun turrets, variable-pitch propellers,[105] retractable undercarriages and more complex radio and electrical equipment, aircraft production was becoming increasingly subdivided and the output of each section of the industry had to be carefully scheduled to prevent shortages of particular components.

During 1936 and 1937, shadow factories were approved for the production of propellers, guns, and carburettors. The most important of these was a factory managed by De Havilland at Lostock which ultimately produced more than half of all the propellers made in Britain during the Second World War. The Air Ministry also persuaded Bristol and Rolls-Royce jointly to establish a new company, Rotol, for the development of variable-pitch propellers.[106] In spite of these measures, however, nothing like sufficient capacity existed in April 1938 to meet the requirements of scheme L, let alone those of the War Potential programme, and with hindsight Lemon was to argue in 1939 that there had been 'a lack of balance . . . which, unless adjusted, would sooner or later have led to a breakdown'.[107]

In planning for both peacetime and wartime production Lemon's staff sought to improve this balance. Between the summer of 1938 and December 1939 the Supply Committee sanctioned capital expenditure exceeding £70 million, which was intended to give Britain the capacity to fulfil the War Potential programme of 2000 aircraft per month by the end of 1941. Apart from airframe capacity, three new factories for Rolls-Royce aero-engines were sanctioned, along with the second Bristol engine shadow group and new factories for both Bristol and Napier. Two carburettor shadow factories were authorised, capacity for Rotol airscrews was expanded, and Lucas, Daimler and Brockhouse Engineering joined Parnall Aircraft and Boulton Paul in the manufacture of gun turrets.[108]

A Directorate of War Production Planning was added to the Air Ministry's new co-ordinating machinery. This drew up a series of war programmes during 1939. The programmes were based on existing plans for the expansion of industrial capacity and on the acceleration believed possible under wartime conditions. Airframe production was placed at the centre of the programme, and requirements for materials, engines, guns, turrets and other equipment were then analysed in detail.[109] Of course the balance was not perfect but the gaps were more the consequence of miscalculation than neglect. The two most serious underestimates, heavy bomber undercarriages and Rotol airscrews, resulted from uncertainty about the extent of wartime requirements for new designs and unexpected difficulties in expanding wartime output.[110] Conversely, an anticipated shortage of aero-engines never materialised.[111]

Space prohibits a detailed analysis of the Air Ministry's activities in each industrial sector, but the depth and scope of the expansion plans for one particular sector, the aluminium industry, may be taken as representative of the Ministry's plans for the aircraft industry as a whole. By the late 1930s, most military aircraft were constructed of aluminium alloys. In the first years of rearmament there were few clear signs of shortages in the supply of this material. The DAP lacked sufficient resources to undertake a detailed investigation of aluminium capacity but supplies appeared to be sufficient for the first-line targets set by scheme F and the Air Ministry had no authority to plan additional capacity for war potential. In any case, a large-scale expansion of the aluminium alloy industry was in progress; between 1936 and 1939 the annual British output of fabricated alloys increased from 6000 tons to 40,000 tons.[112] Early in 1938, however, the aluminium producer ICI (Metals) Ltd produced a survey of British aluminium production suggesting the probable emergence of 'a very awkward hiatus', and the company recommended that shadow factories should be built to augment output.[113] Similar conclusions were reached by Lemon's staff. Adequate capacity existed in Britain to support scheme L, but this 'would be only just sufficient to meet requirements in an emergency if both domestic use and export of aluminium were entirely suspended.'[114]

After the Czechoslovak crisis in September and the acceleration of scheme L, immediate deficiencies were surmounted by importing supplies from abroad, and factory extensions were sanctioned with the objective of increasing the output of aluminium sheet and strip from about 20,000 tons to 30,000 tons per year. Then, as the international situation deteriorated further in the spring of 1939, the decision was taken to double the capacity of the entire light alloy industry. A third of the new capacity was sanctioned in April, and the remainder was put in hand between July and August. Following the finalisation of war production plans in October, seven more projects were authorised together with the purchase of specialised plant.[115]

A matter of increasing concern was the supply of light alloy extrusions, the use of which had become widespread in aircraft design since 1935. During 1939 the Air Ministry made arrangements to double Britain's extrusion capacity but, on the outbreak of war, just 21 extrusion presses were in operation, seven days a week, 24 hours a day, most of which had been made in Germany. Lemon therefore took steps to accelerate press manufacture in Britain by recruiting a suitably experienced engineer to act as adviser to British machine firms. The Ministry scoured the international market and prevented the export of two British presses. Aluminium firms were encouraged to adopt alternatives to extrusions for the production of light alloy sheet and aircraft designers were instructed to use only standard extrusions. Extrusion speeds three times as great as those previously thought possible were achieved after alterations were made to the chemical composition of light alloy.[116]

Another significant response to the threat of light alloy shortages came at the administrative level. In January 1939, a committee of industrialists and Air Ministry officials was established to supervise light alloy distribution and, following the outbreak of war, the light alloy industry set up a voluntary control which organised priority and planning arrangements for the four light alloy sectors: sheet and strip, extrusions, castings, and forgings. The controllers, in turn, co-operated closely with the Air Ministry's Directorate of Materials Production, the aircraft industry and the Aluminium Control of the Ministry of Supply.[117]

The Air Ministry believed that British alloy manufacturers would be able to meet virtually all the requirements of the aircraft programme when the various expansion plans were complete. However, British output of virgin aluminium amounted to only one-quarter of estimated requirements for the first year of the war. By September 1939 the Air Ministry had initiated plans to raise British production from 31,000 tons to 41,000 tons but this still left a substantial deficiency. Limited supplies were expected from the USA, Switzerland and Norway, but the most promising source was the Canadian aluminium industry, which had a capacity of approximately 100,000 tons per year. In September 1939 arrangements were made to import 60,000 tons of virgin aluminium from Canada in the following 12 months. Total production and imports would then amount to 128,000 tons. By mixing alloy scrap with virgin aluminium it was expected to produce a further 32,000 tons, so meeting the total requirement of 160,000 tons.[118] Even combined British and Canadian aluminium production was, however, expected to be inadequate for the second year of a war, so in October the decision was taken to create additional capacity in Canada capable of producing another 100,000 tons per year.[119] By 1941, the Canadian industry was providing Britain with more than half its total aluminium requirements.[120]

These attempts to balance the output of the different sectors of the aircraft industry marked an important development in the progress of industrial planning. However, the new airframe factories, which were last in the sequence of manufacture, were often the first to be sanctioned; many of the plans for increasing the output of ancillary items were initiated later.[121] Moreover, much of the capacity sanctioned between 1938 and 1939 was essentially 'war potential' and was not scheduled to become productive until 1940 and 1941. Of proposals involving an expenditure of over £30 million on capacity for ancillary equipment, Lemon could record as late as July 1939 that only 'some of these have already been approved by the Treasury.'[122]

CONCLUSION

British air rearmament between 1935 and 1938 involved only a limited state aircraft production policy. By encouraging the expansion of the professional aircraft industry the Air Ministry rightly expected to enlarge Britain's war potential. In the selection of aircraft and the allocation of contracts, however, the Ministry was not in a position to introduce what Lord Weir termed a 'war regime' involving group schemes and subcontracting.[123] The immediate aim was to fulfil the first-line targets of schemes C and F. It proved impossible to restrict orders to one type per class because individual firms often lacked sufficient capacity to fulfil all the Air Ministry's demands by the April 1939 deadline and because the Ministry initially refused to contemplate the risks and delays involved in farming out unstandardised designs. The rapid advance of aeronautical technology also militated against any attempt to concentrate available resources on proven aircraft because the proven aircraft of 1936 were invariably obsolete by 1939.

The truth was that a production policy could only develop within limits imposed by industrial factors over which the Air Ministry initially had minimal control. The fundamental constraint on aircraft production was technical change. Until the new technology had been mastered by the firms there was very little that the Ministry could have done to accelerate output. For the same reason the functions of the Air Ministry's production department were restricted to providing advice to the industry and to reporting particular problems to the Air Council.

The development of a more comprehensive state production policy in 1938 was primarily a response to the *Anschluss*; but it was not simply the political situation which had changed. Aircraft ordered at the beginning of rearmament were now entering production. A degree of technical standardisation was attained which had hitherto been lacking. The contractors now had more technical and manufacturing experience than in 1936. At the beginning of rearmament quantity had necessarily been sacrificed to quality. After 1938, output could be multiplied without jeopardising the operational efficiency of the RAF first line.

To implement the new policy, production experts were recruited from the engineering industry. A formal consultative process was established with the industrialists. Separate departments were set up to deal with all aspects of the DGP's work and their functions were co-ordinated through committee meetings and production programmes. Within a short space of time an enormous volume of capital expenditure was approved. Group production schemes were organised; the aircraft industry and its 'shadows' were expanded and more general engineering firms introduced to aircraft manufacturing; subcontracting was adopted as official policy, and much more attention was paid to the supply of aluminium and aircraft equipment. It is true that there were several important oversights. Nevertheless, by September 1939 an industrial structure was emerging which conformed broadly to plans first drawn up by the Air Ministry in the mid-1920s.

While the task of planning war production proceeded, the output of aircraft expanded by no less than 180 per cent during 1939.[124] Intelligence assessments indicated that the output of the British aircraft industry was overtaking that of its German counterpart and this helps to account for the government's willingness to go to war in September.[125] Freeman and Lemon would almost certainly have preferred to wait until the following year when much of the capacity laid down since 1938 was scheduled to commence production. 'It is inescapable', Lemon warned, 'that our war effort, if planned on a long range basis, must entail acceptance of an output curve in the meantime related to pre-war planning.'[126]

The government, however, was unable to accept the implications of such a delay. Rearmament was threatening the country with bankruptcy. In 1939 production expenditure on aircraft and associated items alone totalled £147 million.[127] The value of aircraft contracts placed in that year exceeded £400 million.[128] Such an outlay, when added to the other costs of rearmament, was far greater than the peacetime economy could sustain.[129] Mobilisation was thus an essential prerequisite to the fulfilment of the rearmament and War Potential programmes launched after the *Anschluss*.[130] The challenge facing the Air Ministry in the first months of the war was that of bridging the gap between the achievements of 1939 and war preparations which could not become fully effective until 1941.

NOTES

1. PRO AVIA 46/268, statement of value of aircraft contracts placed, 1924–41.
2. J.A. Cross, *Lord Swinton* (Oxford, 1982), p. 158; PRO AIR 6/24, S of S EPM, 25 February 1936.
3. PRO AIR 2/2014, DAP to AMSO, 10 January 1938; PRO AVIA 10/267, table by DDG Stats. P, 20 December 1941.
4. There were, for example, doubts regarding the supply of aluminium; see PRO AIR 6/33, S of S EPM, 15 February 1938, 22 March 1938, 5 April 1938.
5. PRO AIR 6/58, memorandum by DGP, 12 July 1939; Postan, *War Production*, pp. 66–7.
6. Weir 19/5, interim report of the Air Parity Subcommittee of the Ministerial Committee on Defence Requirements, 8 May 1935.
7. Air Force List.
8. Cross, *Swinton*, p. 156.
9. Reader, *Weir*, pp. 200–2.
10. Ibid., p. 200. According to one contemporary, Cunliffe-Lister had 'no faith except in Lord Weir'. See PRO AVIA 46/210, Air Chief Marshal Sir Wilfrid Freeman (then Chief Executive, Ministry of Aircraft Production) to Postan, 29 September 1943.
11. Reader, *Weir*, p. 201.
12. Weir 19/5, interim report of the Air Parity Subcommittee of the Ministerial Committee on Defence Requirements, 8 May 1935.
13. Weir 19/20, meeting between Weir and McLean, 6 February 1936.
14. PRO AIR 6/58, memorandum by DGP, 12 July 1939.
15. AC 70/10/54, Handley Page to Allen, 2 November 1935.
16. PRO AIR 6/43, meeting of the Air Ministry's Advisory Committee on Contracts, 24 October 1935; PRO AIR 6/23, S of S EPM, 29 October 1935; Postan, *War Production*, p. 450. Progress payments amounting to 80 per cent of estimated material costs were made throughout the inter-war years.
17. Ashworth, *Contracts*, pp. 199–200.
18. PRO AIR 6/43, report by Clegg, 9 October 1935; note by AMSO, 26 October 1935; PRO AIR 6/23, S of S EPM, 29 October 1935.
19. PRO AIR 6/24, S of S EPM, 18 February 1936; Weir 19/2, Weir to AMSO, 27 February 1936.
20. The DAP was not formally brought into the selection process until 1939. See PRO AVIA 15/2364, DAP to DTD, 14 February 1939.
21. PRO AIR 19/9, DAP to AMSO, 27 April 1936; DAP to Weir, 26 May 1936, 3 June 1936, 11 June 1936.
22. PRO AIR 2/1951, PS to AMSO to DAP, 2 March 1937.
23. PRO AVIA 46/91, R. Abrahams' notes on pre-war expansion. Abrahams was Private Secretary to the AMSO.
24. For the DAP's estimates see PRO AIR 2/1790; for actual delivery figures see PRO AIR 19/524.
25. Weir 19/1, DAP to Weir, 2 July 1937; PRO AIR 6/52, Bristol form 1407 (a fortnightly progress report submitted by the firms to the Air Ministry) lists 28 Blenheims in final assembly or erection jigs at any one time in January 1938. Disney's predicted monthly output was 28. See PRO AIR 6/30, S of S EPM, 8 June 1937.
26. A. Harvey-Bailey, *Rolls-Royce: Hives, the Quiet Tiger* (Sir Henry Royce Memorial Foundation, 1985), p. 31. Both Harvey-Bailey and his father were senior technical staff at Rolls-Royce.
27. PRO AIR 2/1908, summary by DTD of experimental projects in progress, 21 October 1936.
28. The Supermarine Spitfire provides the best illustration but design errors also delayed the Battle, the Hurricane and the Blackburn Skua. See PRO AIR 6/31, S of S EPM, 28 September 1937; PRO AIR 6/33, S of S EPM, 22 February 1938; PRO AVIA 46/114, notes on technical difficulties based on information supplied to the Official Historian during his visit to Hawker in 1943; PRO AIR 6/51, note by AMRD, 3 December 1937.
29. PRO AIR 6/51, note by AMRD, 3 December 1937.

30. PRO AIR 2/1790, DAP to AMSO, 11 May 1936, 27 August 1937.
31. AC 70/10/55, Bowyer to Handley Page, 8 September 1937; Weir 19/2, DAP to Weir, 14 April 1937 ('I trust no one in the industry'); RRA, Ernest Hives file, 'War Emergency – 1935–39', Hives (works director at Rolls-Royce) to Sir Arthur Sidgreaves (managing director at Rolls-Royce), 23 April 1938.
32. PRO AVIA 46/91, R. Abrahams's notes on pre-war expansion.
33. PRO AIR 6/32, S of S EPM, 30 November 1937.
34. Weir 19/13, Weir to S of S, 13 December 1937.
35. PRO AIR 6/33, S of S EPM, 29 March 1938; PRO AVIA 10/151, ACCS meeting, 26 May 1938.
36. PRO AIR 6/32, S of S EPM, 30 November 1937, 7 December 1937.
37. Weir 19/13, Weir to S of S, 13 December 1937.
38. Postan, *War Production*, p. 20.
39. Cross, *Swinton*, p. 211.
40. PRO AVIA 46/210, Freeman to Postan, 29 September 1943.
41. Postan, *War Production*, p. 84; W. Ashworth, *Contracts*, p. 202.
42. PRO AIR 6/53, Bruce-Gardner to S of S, 20 April 1938; for minutes of early meetings of the Supply Committee see PRO AVIA 10/151.
43. *Dictionary of National Biography*.
44. AC 70/10/67, file on meetings with the Air Ministry to discuss design and supply; PRO AIR 6/51, note by AMRD, 3 December 1937.
45. Postan, *War Production*, p. 20.
46. *Dictionary of National Biography*; PRO AVIA 46/93, memorandum by Bruce-Gardner, 30 December 1938.
47. PRO AIR 6/58, memorandum by DGP, 12 July 1939; PRO AVIA 10/310, DGP meeting, 17 April 1939.
48. Weir 19/2, Weir to S of S, 21 February 1936.
49. Weir 19/2, the Secretary to S of S, 7 January 1936. Underlined in original.
50. PRO AIR 6/24, S of S EPM, 6 February 1936.
51. PRO AIR 6/26, S of S EPM, 7 July 1936, 21 July 1936.
52. Ibid.
53. PRO AVIA 10/155, second DUS to PDDC, 29 October 1938.
54. PRO AIR 6/55, memorandum by the ACCS, 2 December 1938.
55. PRO AVIA 10/217, ACCS meetings, 5 October 1938, 6 October 1938; Postan, Hay and Scott, *Design and Development*, pp. 10–11.
56. PRO AVIA 46/93, AMDP to Bruce-Gardner, 4 November 1939; Bruce-Gardner to AMDP, 5 November 1939.
57. Ibid.
58. R.J. Overy, *Goering: The Iron Man* (London, 1984), pp. 193–6.
59. Postan, Hay and Scott, *Design and Development*, pp. 92–4.
60. Vickers 410, Vickers-Armstrong (Southampton) daily works report, 23 October 1939.
61. PRO AVIA 15/69, AMDP to DGRD and DGP, 27 January 1940.
62. Tuffen and Tagg, *Hurricane*, p. 50.
63. Postan, Hay and Scott, *Design and Development*, pp. 127–8, 164.
64. Examples include the Blackburn Botha torpedo-bomber and the Saunders-Roe Lerwick flying boat. The resulting deficiencies were made up by adapting other designs (like the Bristol Beaufighter), by maintaining older models in production (like the Fairey Swordfish and the Short Sunderland), and by importing aircraft from America.
65. Hornby, *Factories*, pp. 395, 402.
66. Weir 19/1, Secretary to S of S, 7 January 1936.
67. Ibid.
68. This point was stressed by Swinton during the Air Council's discussion on selecting the Blenheim for shadow production. See PRO AIR 6/24, S of S EPM, 28 February 1936, 3 March 1936.
69. Ernest Hives of Rolls-Royce believed shadow aero-engines would always be obsolete by the time they entered production. See RRA, Hives file, 'Policy', Hives to Sidgreaves, 13 September 1937.
70. R.J. Overy, *The Air War* (New York, 1981), p. 212.
71. For the relevant production figures see PRO AIR 19/524.

72. PRO AIR 19/5, Herington (Secretary, Aero-Engine Shadow Committee) to S of S, 21 April 1936; AMSO to S of S, 20 April 1936.
73. PRO AIR 19/9, DAP to Weir, 26 May 1936, 3 June 1936, 11 June 1936.
74. *Aircraft Production*, November 1938.
75. £450,000 was spent on jigs and tools for an order of 863 airframes: £521 per airframe. See PRO AIR 6/37, S of S EPM, 24 January 1939.
76. PRO AVIA 10/217, ACCS meetings, 5 and 6 October 1938.
77. PRO AIR 6/55, memorandum by the ACCS, 2 December 1938.
78. AC 72/3, report by manager of Bristol Shadow Industry Office, 'History of Production of Bristol Aero-Engines by the Shadow Industry Operating under the Joint Aero-engine Committee in Conjunction with the Bristol Aeroplane Company Limited', pp. 17–18.
79. PRO AVIA 10/12, meeting between the Air Ministry, Rolls-Royce and Ford, 31 October 1939.
80. RRA, Hives file 'Policy', Db to Hives, 14 September 1937.
81. Overy, *Air War*, p. 248.
82. PRO AIR 6/43, DS to the Secretary, 23 November 1935.
83. PRO AIR 8/196, DRC minutes, 20 May 1935.
84. PRO AIR 6/43, the Secretary to S of S, 23 July 1935. Embodiment loan equipment comprised the bulk of those aircraft components (propellers, engines, armament and radio equipment) which were not manufactured by the airframe industry. This equipment was purchased by the Air Ministry and issued free to the airframe firms for installation.
85. Weir 19/5, interim report of the Air Parity Subcommittee of the DRC, 8 May 1935.
86. PRO AIR 2/714, meeting between the Air Ministry and Hawker, 6 November 1934; DDTD to D of C, 20 November 1934.
87. PRO AIR 8/196, DRC minutes, 20 May 1935.
88. Weir 19/2, Weir to the Secretary, 1 August 1935; PRO AIR 6/43, the Secretary to S of S, 23 July 1935.
89. PRO AIR 6/51, note by AMRD, 3 December 1937.
90. PRO AIR 2/3273, DAP to AMSO, 12 April 1938; PRO AIR 2/2711, DAP to AMSO, 24 May 1938.
91. PRO AIR 6/33, S of S EPM, 22 March 1938.
92. Weir 19/23, notes by Weir for a meeting with the Secretary of State for Air and the Minister for Co-ordination of Defence, 15 March 1938.
93. R.A.C. Parker, 'British Rearmament 1936–39: Treasury, trade unions and skilled labour', *English Historical Review*, 96 (1981), p. 337.
94. PRO AVIA 10/151, note by AMRD, 14 May 1938.
95. PRO AIR 6/54, memorandum by DGP, 5 September 1938; PRO AIR 6/35, S of S EPM, 14 September 1938.
96. Parker, 'British Rearmament', pp. 338-9; PRO AVIA 10/154, ACCS meeting, 28 July 1938.
97. PRO AIR 6/54, memorandum by DGP, 5 September 1938.
98. PRO AVIA 46/93, the Secretary to Bruce-Gardner, 13 September 1938.
99. PRO AIR 6/58, memorandum by DGP, 12 July 1939.
100. Hornby, *Factories*, p. 227.
101. PRO AVIA 46/91, Ord to Postan, 15 July 1943.
102. PRO AVIA 10/243, 'The Application of Wage Incentive in the Aircraft Industry to Facilitate Cost Control', by T.S. Smith (undated but late 1939); J.V. Connolly papers, 'Survey of the Relationship between Labour Effort, Aircraft Size, Jig and Tool Cost, and Rate of Output', by J.V. Connolly, 15 January 1940. J.V. Connolly joined the Air Ministry's Directorate of Statistics and Planning in September 1939.
103. AC 70/10/79, DGP to Handley Page, 14 December 1938.
104. PRO AIR 6/58, memorandum by DGP, 12 July 1939.
105. The angle of the blades to the airstream can be varied in flight with a variable pitch propeller.
106. Hornby, *Factories*, pp. 202-3; PRO CAB 102/47, 'Development and Production of Propellers', by D. McKenna (unpublished official narrative), pp. 17–18.
107. PRO AIR 6/58, memorandum by DGP, 12 July 1939.

108. Hornby, *Factories*, pp. 202–3, 255–75.
109. PRO AIR 6/58, memorandum by DGP, 12 July 1939.
110. Hornby, *Factories*, pp. 269, 272–3.
111. PRO CAB 102/51, sections 91–3.
112. PRO AVIA 46/72, memorandum by the Air Ministry prepared for the Select Committee on National Expenditure, 28 March 1940.
113. PRO AIR 6/33, S of S EPM, 15 February 1938. ICI (Metals) Ltd was at that time a leading manufacturer of fabricated light alloys.
114. PRO AVIA 10/154, ACCS meeting, 5 August 1938.
115. PRO AVIA 46/72, DGP to S of S, 28 March 1940; memorandum by the Air Ministry for the Select Committee on National Expenditure, 28 March 1940; PRO AIR 6/58, memorandum by DGP, 12 July 1939.
116. PRO AVIA 46/72, DGP to S of S, 28 March 1940.
117. PRO AVIA 10/267, memorandum by the Air Ministry (undated but early 1940), 'Fabricated Light Alloys (Aluminium and Magnesium) – Production and Distribution'; meeting of the joint committee of the Light Alloy Controllers and the SBAC, 12 February 1940.
118. PRO CAB 102/187, 'Aluminium and Aluminium Alloys, 1939–45', by M.E. Jenkins (unpublished official narrative), pp. 23–4, 31; PRO AVIA 10/102, review by the Air Ministry of the raw material position at 28 September 1939.
119. PRO AVIA 10/267, meeting between the ACCS and Mr Cunliffe (Ministry of Supply aluminium controller), 19 October 1939. The total additional capacity created eventually amounted to 107,500 tons per year. See PRO CAB 102/187, p. 28.
120. Hornby, *Factories*, p. 279.
121. Ibid, p. 202.
122. PRO AIR 6/58, memorandum by DGP, 12 July 1939.
123. Weir 19/23, notes by Weir for a meeting with the Secretary of State and the Minister for Co-ordination of Defence, 15 March 1938.
124. *Statistical Digest of the War* (London, 1951), p. 153.
125. P.M.H. Bell, *The Origins of the Second World War in Europe* (London, 1986), p. 181.
126. PRO AVIA 46/72, DGP to S of S, 24 January 1940.
127. Postan, *War Production*, p. 172.
128. PRO AVIA 46/268, statement of value of aircraft contracts placed, 1924–41.
129. Shay, *British Rearmament*, pp. 276–80.
130. R.J. Overy with A. Wheatcroft, *The Road to War* (London, 1989), pp. 99–100.

3

The airframe industry and military aircraft production, 1935–39

The main features of the British airframe industry in the inter-war years were described in Chapter 1. By the standards of the day the industry was large and its output was relatively high. It was dominated by three companies: Hawker Siddeley, Vickers, and De Havilland, but its capacity was far lower than that considered necessary by the government to fulfil wartime requirements. Moreover, the emphasis on military and private aircraft design in Britain had resulted in a slower transition from the biplane to the monoplane than that achieved in the United States and Germany and this transition was far from complete at the beginning of rearmament.

The Air Ministry harboured no illusions about the qualitative inferiority of British aircraft compared with the best foreign types in 1935. At the beginning of scheme C, Lord Weir denounced what he saw as 'a deplorably weak situation'. While several promising monoplanes were under development, such aircraft did 'not constitute any part of the definite programme'. In the medium bomber class 'no really modern type whatever [was] available for production'. Light bomber requirements were to be met by obsolete biplanes like the Hawker Hind. 'By March, 1937', Weir concluded, 'our 30 light bomber squadrons, three sevenths of our total strength, will be machines with a cruising speed of 150 mph, carrying 500 lbs of bombs at 450 miles range.'[1]

Substantial numbers of obsolescent types were constructed between 1935 and 1938. In 1936, for example, Hawker biplanes alone accounted for 50 per cent of sales to the Air Ministry and, even in the following year, the figure was as high as 29 per cent.[2] Given the prevalence of such aircraft in first-line service in the mid-1930s, the Air Staff's eagerness to re-equip the RAF with more modern designs during the course of scheme F is easily understood and there was initially considerable optimism in both the Air Ministry and the aircraft industry that the programme could be achieved by the target date of 1 April 1939. 'The production of a total of 655 machines by March 1939 will be easily attainable', Richard Fairey told Newall, 'and represents a low normal for a factory the size of Stockport.'[3] In May 1936 Disney noted that 'great strides have been made . . . with [Vickers'] geodetic system'. It was thought that the first Wellingtons would emerge in May 1938 and that production might eventually reach one per day. Disney also saw Bristol as 'a most efficient production unit . . . capable of producing all the medium bombers we want of them'.[4]

These were indeed confident predictions. The acceleration of rearmament in 1936 forced some companies to re-jig for large-scale production and this alone was the cause of some delay, yet the industry's task was not simply physical expansion, nor an increase in output. Although the expansion of capacity took time, it is quite possible that the original production targets would have been achieved if rearmament had been purely a question of quantity. Instead the aircraft industry faced the much more formidable challenge of co-ordinating expansion with the adoption of a novel and highly sophisticated technology. The new monoplane bombers and fighters were both larger and much more advanced than any aircraft previously built in Britain; their design and development caused serious technical problems which took longer to solve than the firms initially expected. Until 1938, deliveries of these types fell far short of the levels promised by the industry or anticipated by the government, and relations between the two parties deteriorated drastically as a result.

The Air Ministry's stock explanation for the problems of the early rearmament years was that the firms were inefficient. Several historians have accepted the Ministry's view, and the charge would certainly have been justified had the industry proved incapable of producing standardised designs in quantity. Given the absence of standardisation between 1936 and 1938, however, allegations of inefficiency are unwarranted. Technical change, more than any other factor, prevented the industry from fulfilling its production goals and, in so far as technical change was necessary to improve the operational quality of first-line aircraft, it represented an obstacle which was impossible to avoid.

One aim of this chapter is to illustrate in detail some of the ways in which the revolution in aeronautical technology disrupted British aircraft production. As soon as a degree of design standardisation was attained, however, the industry broke free of these technical constraints. Production accelerated rapidly in the 12 months preceding the outbreak of war, reaching higher levels than originally expected by the Air Ministry.

PRODUCTION DELAYED

On paper, scheme F envisaged the production of 8000 aircraft between March 1936 and March 1939. In practice, some carry-over beyond the deadline date was always expected. In programmes drawn up for 3000 of the more modern types, for example, the Air Ministry calculated that 600, or one-fifth, would be delivered after March 1939. Nevertheless, production in 1937 and 1938 was far lower than originally anticipated. In 1937, 220 Fairey Battles were originally expected; actual output totalled only 80. Of the 330 Bristol Blenheims scheduled, only 117 were delivered. Spitfire production was expected to start in September 1937, and 165 were scheduled for delivery in the following twelve months. The RAF received just five. It is difficult to assess the total deficit because the Air Ministry's plans were subject to continuous revision, but programmes for 12 of the 15 principal airframe firms provided for the production of 2200 aircraft in 1937. In the event this was as much as the entire industry could deliver in that year, and the proportion of obsolescent types was considerably higher than expected.[5]

The weaknesses of the Air Ministry's programming machinery have already been discussed. When these schedules were drawn up Disney himself warned that 'it is still too early to give a reliable forecast' and that 'the figures on the programme are pure guesswork'.[6] Yet the programmes were prepared in consultation with the firms, and both sides apparently believed that they were not unduly optimistic. Why, then, was production so seriously delayed between 1936 and 1938?

The transition from the relatively limited aims of scheme C to the much more ambitious targets of scheme F had far-reaching industrial implications, both for the aircraft industry and for the engineering industry as a whole. Production had to be extensively reorganised, and reorganisation took time. Under the earlier scheme, contracts for the new monoplanes never exceeded 150 aircraft and for most firms were considerably smaller.[7] The much

TABLE 7

Bristol Production Costs

Cumulative Aircraft	Production Cost Units	
	Grade A Tools	Lower Grade Tools
50	175	70
200	50	35
400	34	35
800	25	35

Source: Connolly Papers, data supplied by H.J. Pollard of the Bristol Aeroplane Company.

larger orders placed for scheme F compelled the industry to invest far greater sums in plant and machinery, as the SBAC recorded in 1937:

Plans which were adequate for jigging and tooling production of, say, less than one hundred aircraft were unsuited to the production of several hundreds. Plans drawn up to meet the requirements of the 1935 pro- gramme and already partially fulfilled had to be scrapped and fresh plans prepared.[8]

This was primarily a question of economics. For example, in a comparison (see Table 7) between tooling costs and manufacturing costs, Bristol found that the use of the most elaborate and expensive ('grade A') tools could not be justified until 400 units had been produced. Thereafter the employment of such tools resulted in a continuous fall in manufacturing costs, while the use of less sophisticated tooling brought these reductions to a halt. Data for the Spitfire (see Table 8) indicates that a similar 'break even' point

TABLE 8

Spitfire Detail Part Production Costs

Cumulative Aircraft	Production Cost (pence)	
	Press Tools	Bench Tools
100	385	225
300	133	100
500	82	75
1000	44	57

Source: PRO BT 28/423.

could not be reached until at least 500 units had been produced with the most advanced tools ('press tools').[9]

So when, in May 1935, the EEF suggested that aircraft firms should make more use of press tools, it was pointed out that the quantities of aircraft on order were 'frequently too small to make this a practicable proposition from the point of view of cost'.[10] In February 1936, however, Fairey initiated very extensive tooling arrangements for the Battle and, in the same month, Bristol reported to the Air Ministry that 'in view of the materially increased programme we are now completely reconsidering our production tooling'.[11]

The level of tooling was also influenced by the availability of labour. Jigging and tooling allowed firms to increase the proportion of semi-skilled labour employed on the production line but the tools themselves had to be made by skilled tool-makers who were in short supply. The industry was reluctant to aggravate this shortage unless there were overwhelming arguments in favour of engaging larger numbers of semi-skilled workers, and during scheme C most firms decided to tool up on a limited scale and to retain a high proportion of skilled labour for production.[12] Scheme F, however, dictated the opposite policy to the largest contractors because there was insufficient skilled labour available to achieve the increased production targets. By the early months of 1936, therefore, considerations of both cost and manpower were forcing the industry to accept a much higher level of mechanisation than had previously been employed. The revision of earlier plans tended to disorganise production and took longer to complete than originally expected because of delays in the delivery of jigs and tools sometimes amounting to more than 12 months.[13]

This was a complex problem and not merely a reflection of inadequate domestic capacity for small tool manufacturing. Jig and tool manufacture could not even begin before new aircraft designs had been standardised. Moreover, deliveries would have been greatly improved had the number of different aircraft in the Air Ministry's programme been reduced. Every different type of aircraft required a different set of jigs and tools; a more rational aircraft programme would have allowed for much more efficient jig and tool production.[14] Yet there were very good reasons for the variety of designs and, under the circumstances, there was no alternative but to accept that conditions were not ideal for maximising output.

The constraints imposed by purely physical shortages are, in any case, easily exaggerated. The key obstacle facing the aircraft industry was the technological revolution of the mid 1930s. The externally braced and strutted fabric-covered biplanes which had served the RAF for much of the inter-war period had now to be replaced by cantilever (usually) metal-skinned monoplanes, embodying devices such as variable-pitch propellers, retractable undercarriages and tail wheels, and wing flaps. As the aerodynamic efficiency and performance of aircraft increased, so too did the loads and stresses imposed on aircraft structures. At the same time military demands became much more severe.

Mastering this new technology was always certain to be a long and complex process and the aircraft industry was not entirely to blame for the difficulties that arose. When apportioning work between the firms, the Air Ministry took no account of design capacity and, consequently, available resources were dispersed among too many different projects. Neither Hawker nor Blackburn controlled more technical facilities than the majority of companies, yet between them they received design contracts for no fewer than seven different types in 1936.[15] At Hawker the diversion of effort to unsuccessful aircraft like the P4/34 bomber and the Hotspur fighter delayed metal wing development for the Hurricane.[16]

Most firms had to design and produce at least two different aircraft during the course of rearmament. Vickers built the Wellesley and the Wellington, Handley Page built the Harrow and the Hampden, and Gloster built the Gauntlet and the Gladiator. The production of the Wellesley at Vickers was set back by three months because resources were diverted to the Wellington prototype.[17] Handley Page criticised as 'wasteful' the Air Ministry's decision to order only 100 Harrows. The technical expenditure involved was high in relation to the number of aircraft actually built and production ceased just as peak output was reached.[18] Gloster's managing director believed that output could have been increased by a third in 1936 and 1937 had his company been able to concentrate on a single type.[19]

Service demands for ancillary equipment also caused problems for the industry. One design, De Havilland's Don trainer, failed completely because of excessively complex RAF specifications. As Lord Swinton commented, 'the amount of equipment specified . . . was more than it was reasonable to expect the aircraft to carry'.[20] Indeed the challenge facing the industry lay not only in mastering a new technology but in accommodating military requirements within that technology. 'Britain First', the all-metal civil monoplane built by Bristol for Lord Rothermere, flew in 1935, but the aircraft had to be almost completely redesigned before it emerged in 1937 as the Blenheim bomber.[21]

The Air Ministry recognised that it bore some responsibility for guiding new projects through design and development and into production. As already noted, proposals were periodically made to reduce the Ministry's influence over design. This became the official policy in 1934 but it is impossible to know how effective it really was. Complaints from contractors about unwarranted interference by government 'experts' remained common throughout the rearmament years and in December 1937 the AMRD complained, just as the DTD had complained in 1934, that 'specifications are . . . far too detailed'. He went on to suggest that 'specifications should be simplified to the extent of concentrating on the fundamental requirements, leaving more latitude to designers than at present'.[22]

New aeronautical and military technology created fundamental difficulties for the aircraft firms. These recurred throughout the design history of the majority of aircraft built in the mid-1930s, but Handley Page's Hampden bomber provides the one of the best illustrations. Handley Page was regarded as a highly competent manufacturing unit by the Air Ministry and, during the rearmament years, the company initiated several important changes in aircraft production methods which brought substantial gains in productivity. Such advances, however, counted for little until the fundamental design problems facing the industry had been overcome.

At the commencement of rearmament, Handley Page underestimated the technical effort required to develop the Hampden successfully. More than twice the number of man-hours was absorbed in its design than in that of Handley Page's previous model, although it was apparently hoped that design costs might be reduced.[23] The technical problems were exacerbated by the very low weight limits demanded by the Air Staff's original B9/32 bomber specification, from which both the Hampden and the Vickers Wellington were developed. The removal of these restrictions after the collapse of the Geneva disarmament talks helped persuade both companies to redesign their respective tenders.[24]

The spiralling increase in expenditure which accompanied the development of the Hampden was accepted by Handley Page with the greatest reluctance. As early as December 1934 the firm's chief draughtsman, Ratcliffe, circulated a gloomy memorandum to other senior executives stating: 'We have now spent a considerable sum of money in the drawing office and so far no finished drawings have been produced.' Ratcliffe referred to numerous alterations which had made it almost impossible to estimate design costs accurately. 'This may be a particularly unfortunate design', he concluded, 'but when the day of reckoning arrives, the writer hopes that all concerned will remember these changes and the various reasons for the high design costs of this aircraft.'[25] Only in March of the following year was Ratcliffe able to provide a timetable for the completion of design work and, in doing so, he again warned against making subsequent alterations. 'No departure should be made without the full effect on the design programme being thoroughly recognised.'[26]

Prototype design and manufacture subsequently proceeded in parallel with the Vickers Wellington. Handley Page monitored the Wellington's progress closely, and when its qualitative superiority became apparent there were fears that the entire Hampden project would be abandoned.[27] A host of minor technical troubles delayed the flight of the prototype and although Handley Page ultimately received a production order (for reasons discussed in the previous chapter) flight trials disclosed the need for a number of modifications and it was not until July 1937 that the prototype was finally accepted. Two months later the Air Ministry decided that the armament installation should be changed and as a result the front fuselage had to be completely redesigned.[28]

As a result, some basic design work was still in its initial stages nearly six months after the commencement of production planning and jig and tool design and manufacture. Modifications to the Harrow and preparations for future projects like the Halifax dictated that, at peak, only 69 draughtsmen could be employed on the Hampden. To the first flight of the prototype in 1936 the technical man-hours spent on the aircraft numbered 180,000, but this figure represented only 53 per cent of total technical man-hours. Detailed drawings were not completed until August 1938, the month in which the first production aircraft was delivered to the Air Ministry.[29]

Once mastered, the new technology had to be incorporated into aircraft which were suitable for large-scale manufacture. Within the firms this question caused considerable friction between production and design staff. Designs were often criticised for their excessive complexity, although designers could always argue that it was performance rather than ease of production which ultimately sold military aircraft.[30] It was typical that Vickers' designer, Barnes Wallis, should have complained retrospectively of 'the practical shop element of personnel at [Vickers] Weybridge which was absolutely dominant at the time . . . [and] was bitterly opposed to geodetics', while a former Vickers works manager recalled that geodetic construction was expensive in tooling and difficult in production. Wallis was eventually persuaded to accept design changes in the interests of cheapening manufacture.[31]

This process was also influenced by the time limit which the government set for scheme F. The Air Ministry's primary goal was to establish a particular first-line air-force strength by April 1939; the creation of industrial capacity or war potential was a secondary consideration. Hence in 1936 Richard Fairey readily admitted that the Battle had not been designed for large-scale output but his proposal to redesign the aircraft and to introduce 'a repetition system for really large quantities' was rejected because the Air Ministry believed that initial deliveries would be set back.[32]

The Battle's wartime role was relatively unimportant but the Vickers Wellington was produced in larger numbers than any other British bomber during the following decade. In September 1936 Weir complimented 'Mr Wallis's progress in simplifying the design and his new system of standardising unit parts', but further innovations aimed at making the Wellington easier to produce were, again, not adopted because they would have delayed the introduction of the aircraft. When rearmament was accelerated in 1938 there was no time to incorporate these changes retrospectively.[33] Long-term production economies seemed unimportant in 1936 because only 180 Wellingtons were on order. Had it been known that 11,000 of these aircraft would eventually be produced, the Air Ministry's priorities might have been very different.

The problems of co-ordinating radical changes in technology with large-scale production plans did not differ greatly from firm to firm. The chief difficulty lay in ensuring that the design and production of different components proceeded in parallel. Invariably, particular parts were delayed by technical problems and the early stages of production were thrown out of balance as a result. In May 1936, for example, the records show that Bristol had manufactured 32 Blenheim fuselages but no wings, and only six sets of wings had been completed by November.[34] At the end of March 1938 Supermarine had completed 35 Spitfire fuselages, but only four sets

of wings had been received from the company's subcontractors.[35] At
Hawker, too, the development of stressed-skin metal wings was a
bottleneck. The first Hurricane was designed with a fabric-covered
wing, and 500 such aircraft were delivered to the RAF before the
metal-winged version appeared in March 1939. Throughout 1940
the level of Hurricane production was primarily determined by the
availability of this component.[36]

The Fairey Battle was similarly delayed by technical problems
relating to the engine installation, the bomb carrier and the wings,
and while assembly was held up pending the resolution of outstanding design questions, manufacturing and sub-assembly operations
advanced. By the time the design was cleared in its entirety, these
earlier stages of production were expanding rapidly and it proved
impossible to absorb new labour on assembly work quickly enough
to balance the output of the different sections of the factory.[37] In
June 1937 a member of Disney's staff reported of Fairey's Stockport works that 'there is so much finished and partly finished work
in the shops that unless the aircraft do come out shortly there will be
no space left to work in'.[38]

Until these designs had been standardised in their entirety there
was little that could be done to balance production. Bristol installed
additional wing jigs and subcontracted wing assembly to the Bolton
firm, Dobson and Barlow.[39] Supermarine duplicated wing jigs
and tools, and Sunday working and night shifts were extended to
accelerate wing production.[40] A special section was created within
the Hawker drawing office for metal wing development.[41] Fairey
transferred foremen and skilled labour from their Hayes factory to
Stockport and, under pressure from the Air Ministry, recruited two
new assistant managers.[42]

Despite these and similar measures, deliveries of most of the
new monoplanes started six to nine months later than the dates
envisaged by the Air Ministry's first production programmes. Only
the realisation of higher rates of production than those originally
expected allowed the accumulated deficit to be reduced, and in
some cases eliminated, by the scheme F target date of 1 April 1939
(see Table 9).

TABLE 9

Estimated and Actual Aircraft Deliveries for Scheme F

Firm/Aircraft	Scheme F Target	Deliveries at 1 April 1939	Date Target Achieved
Fairey Battle	526	551	Feb. 1939
Bristol Blenheim	850	667	Aug. 1939
Hawker Hurricane	492	315	Aug. 1939
Vickers Wellington	168	87	Aug. 1939
Supermarine Spitfire	270	130	Jul. 1939

Source: production programmes: PRO AIR 2/1790; delivery charts: PRO AIR 19/524.

STATE-INDUSTRY RELATIONS

The production problems encountered during 1936 and 1937 soured relations between the aircraft industry and the government. The two sides became distrustful of one another as the firms reneged on the promises they had made in 1936 and the Air Ministry brought pressure to bear on the most troublesome contractors. The Ministry concluded that bad management was the fundamental cause of the delays and managerial changes were forced on Fairey, Blackburn, Boulton Paul, Westland, Vickers, Bristol and Airspeed.[43] The smaller firms, like Westland or Boulton Paul, quickly agreed to co-operate but when larger concerns, like Vickers, proved less accommodating, pressure was brought to bear at the very highest levels.

Following delays at both Vickers Aviation and Supermarine late in 1936, Swinton concluded that better results would be achieved if the airframe companies were supervised more closely by their parent concern. Fortunately, a former school friend, Sir Archibald Jamieson, happened to be a senior Vickers director.[44] In February 1937 Weir discussed the aircraft production problems with Sir Charles Craven, chairman and managing director of Vickers' armaments subsidiary, Vickers-Armstrong, stating that the Air Ministry 'had come to the conclusion that the management [of Vickers' airframe subsidiaries] must be improved'. Craven and Swinton then persuaded Jamieson to end the relative independence enjoyed by the airframe firms by turning both companies into subsidiaries of Vickers-Armstrong (see Figure 2). McLean joined the board of Vickers-Armstrong and retained his day-to-day control over the two aviation concerns, but he was compelled to accept Craven's overall management and to agree to a major reorganisation of his senior staff.[45]

In other cases where production fell back seriously, the Air Ministry adopted an even stronger line with the industry's senior executives who, in turn, reacted indignantly to charges of inefficiency. Air Ministry criticism of Fairey's Stockport factory provoked the general manager to threaten Disney with legal action.[46] Subsequent attempts by the Ministry to organise an independent investigation of the factory proved unsuccessful.[47] Richard Fairey accused the Air Ministry of prejudice against his company and refused to admit that there were any weaknesses at the plant other than those which might reasonably be expected following the creation of any new production organisation.[48]

Fairey's resentful tone reflected serious alarm at the extent of Air Ministry interference which was shared by other industrialists. The firms believed that they were being blamed for problems which were often not of their own making, and feared that the government would view controls, or even outright nationalisation, as a means to counter political criticism. The case for nationalisation had been strongly argued by the Left at the beginning of rearmament but even members of the largely Right-wing Air Committee in Parliament felt that the defence programme would benefit from 'the industry either rationalising itself, or being compulsorily rationalised by the government'.[49]

Although the Air Ministry rejected these arguments, the shadow industry was mistakenly seen by some businessmen as a move in the direction of nationalisation,[50] and concern mounted during 1937 as the production schedules slipped back. The Italian aircraft industry had already been nationalised, and its French counterpart was suffering the same fate at the hands of the Popular Front. Critics of rearmament compared the British aircraft industry unfavourably with its German counterpart, which was partly state owned.[51] The industry's fears were summed up in a letter from one senior member of the SBAC staff to Frederick Handley Page in September:

I am most disturbed about reports of the output of new monoplanes for the Royal Air Force. Very disquieting figures of present production have reached me . . . The future of the industry as a private enterprise depends

FIGURE 2
Vickers Ltd Aircraft Organisation

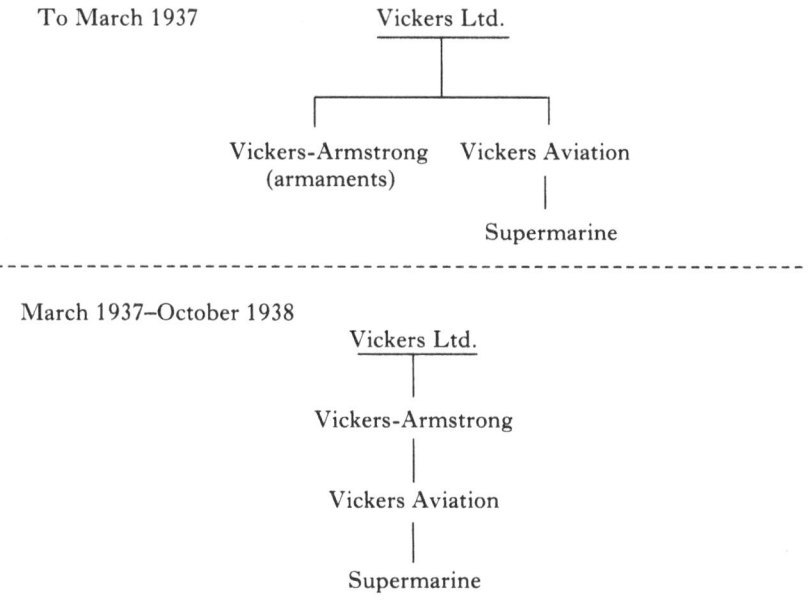

To March 1937

Vickers Ltd.

Vickers-Armstrong   Vickers Aviation
(armaments)

Supermarine

March 1937–October 1938

Vickers Ltd.

Vickers-Armstrong

Vickers Aviation

Supermarine

on our putting forward concrete proposals . . . Taking the industry as a whole, we are apparently not really producing the goods. I see nationalisation very near, to cover up Air Ministry blunders. Anyway, an industry which is absolved from loss and is placed on a 'cost plus x per cent' basis is already nearly nationalised.[52]

The firms were deeply suspicious of Disney's ambitions within the Air Ministry and of group production proposals put forward in 1937. The DAP originally intended to pool the resources of several firms in a Wellington bomber group, which was to be supervised by his staff, but the principle of Air Ministry control was rapidly abandoned after Sir Frank Spriggs of Hawker Siddeley 'stated that he would resist such a scheme to the very limit of his powers . . . to the extent of taking it to the highest quarters'.[53] The industry insisted that parent firms, rather than state officials, should supervise the groups established during the following three years.[54] Some industrialists also feared the implications of direct government investment after the introduction of scheme L. 'With a socialist government in power', Handley Page warned, 'it is but a short step to nationalisation and public utility companies.'[55] This concern re-emerged during the war when the government demanded access to the firms' private accounts.[56]

Private ownership of the aircraft industry was usually justified on the grounds that it was more efficient and produced more rapid technical advance than state ownership.[57] This rationale was undermined by the production problems encountered by the firms in the early rearmament years. In the circumstances it seemed essential to shift the blame to the government, and during the second half of 1937 the SBAC composed a lengthy memorandum which attempted to explain the industry's difficulties and which argued that excessive state interference was one of the principal causes of the delays. Intended for Parliament and the press, it is unclear whether this document was ever circulated but the industry certainly provided material for the parliamentary campaign against the Air Ministry which led to Swinton's resignation in May 1938.[58] At about this time proposals both to rationalise the industry and to reduce official control over design were in fact being made within the Air Ministry.[59]

Two initiatives were taken to improve matters. First, several meetings were held between the two parties at the end of 1937 to discuss design and production problems.[60] Second, the government persuaded the SBAC to appoint as full-time independent chairman the prominent industrialist, Sir Charles Bruce-Gardner. In the latter instance an SBAC delegation was summoned to Downing Street at short notice and apparently without prior knowledge of the government's precise intentions.[61] The delegation, comprising representatives of just three firms, agreed to the new appointment immediately when dissent might have been expected given the prevailing resentment within the industry over government interference and the fact that Bruce-Gardner was a rationalisation expert.[62]

Why did the SBAC agree so readily? There is no clear explanation but it is possible that the industrialists were given little choice. Both Weir and Swinton saw Bruce-Gardner's appointment as an alternative to imposing controls on the industry. Weir actually envisaged telling the SBAC that 'to avoid full-blooded State control we suggest to you that it would be in your best interests for you to look round and find a full-time leader for your industrial association'.[63] There must at least be a suspicion that the Prime Minister's proposal to the aircraft firms was couched in precisely these terms.

Bruce-Gardner's appointment did not at first lead to greater mutual understanding or co-operation. During the joint discussions on design and production the firms consistently complained that they were not receiving sufficient information about the government's long-term intentions and that decision-making within the Air Ministry was slow and excessively bureaucratic. Ample justification for these criticisms was, in their view, provided by the six-week delay between the *Anschluss* and the final approval of scheme L. In the interim, to add insult to injury, the government announced that it was sending an aircraft purchasing mission to the United States.[64]

There were still more grounds for disharmony. At the end of 1937 the Air Ministry made a number of contractual proposals which were less favourable to the industry than terms that had been agreed in 1936.[65] Relations therefore tended to decline even further during the early months of 1938.

Most of the disputes between the Air Ministry and the aircraft industry in the early years of rearmament resulted from the unrealistic promises which the constructors made at the beginning of scheme F and from the absence of qualified industrial planners and statisticians in the Air Ministry. With the advantage of hindsight, however, it can be argued that the deterioration of state–industry relations resulted, in part at least, from two misconceptions.

The first was that production failed to expand during 1937. During 1938 the Air Ministry started to record production by the weight, as well as by the number, of aircraft produced. Had this measurement been adopted earlier it would have been recognised that a very considerable improvement in aircraft production had actually taken place in 1937, for while the number of aircraft delivered only increased by 21 per cent compared with 1936, the output of structure weight increased by 74 per cent.[66] There was also an increase of more than 50 per cent in per capita output (see Table 10). It seems doubtful that a much more rapid expansion could reasonably have been anticipated, and more realistic expectations would certainly have eased tensions between state and industry during this period.

The second misconception was that aircraft ordered in 1936 would be obsolete by the time they entered production. Although this was certainly true in some cases, in others the state ultimately received far better value for money than could possibly have been expected at the beginning of rearmament. Models like the Hampden and the Wellington were larger and better aircraft than those originally ordered and were also capable of carrying heavier loads over longer distances. Moreover, in the case of the Wellington, there existed the potential for much greater development in the longer term.[67] Similarly the Spitfire, an aircraft ordered in small quantities to fill a gap in the 1936–39 fighter programme, remained at the forefront of piston-engined fighter technology throughout the Second World War. From the Air Ministry's viewpoint in 1937, however, these were aircraft which would have a short service life, and the longer their introduction was delayed the shorter that life would be.

TABLE 10

British Military Airframe Deliveries and Employment, 1935–39

Year	Aircraft Delivered	Structure Weight (m.lb)	Labour (June)	Per Capita Output (lb)
1935	893	1.91	15,000	127
1936	1,830	3.75	36,000	104
1937	2,218	6.54	40,000	163
1938	2,828	9.82	65,000	151
1939	7,940	28.89	130,000	222

Source: Labour figures from PRO AIR 8/196 (April 1935) and PRO AVIA 12/194 (1936–39); delivery and structure weight figures from *Statistical Digest of the War*; 1935–37 figures do not include subcontractors.

THE EXPANSION OF AIRCRAFT PRODUCTION

Aircraft deliveries fell slightly during the first half of 1938, but production expressed in structure weight continued to rise. Finally, during the second half of the year, both figures climbed dramatically. Structure weight and aircraft output in the last quarter of 1938 were more than double the level of the first quarter. Deliveries from Fairey's Stockport factory averaged 28 aircraft per month whereas the Air Ministry had originally predicted a peak of 20 per month. Peak monthly deliveries from Bristol amounted in 1938 to 49 aircraft and in 1939 to 61 aircraft; the original programmes envisaged a peak of just 35. The industry's total annual output rose from 2800 in 1938 to nearly 8000 in 1939.[68] Given the disappointments of the early rearmament years and the decline in state–industry relations, the quantitative achievements of British aircraft production between 1938 and 1940 require detailed explanation. Several different factors were involved: investment in fixed capital, better state–industry planning, the stabilisation of design, improved design and production methods, and the extended use of sub-contracting.

The expansion of airframe capacity, 1935–38

The technical problems which delayed the aircraft programme during the early stages of rearmament also tended to conceal very extensive reserves of productive capacity which were underutilised until the later months of 1938. Some firms had spare floor space at their disposal in 1935. Blackburn had vacated a factory in Leeds in 1932, but reoccupied it in 1936.[69] Supermarine's Hythe works was reopened after several years of closure.[70] It is therefore difficult to establish the precise extent of the airframe industry's expansion in terms of floor space between 1935 and 1938. Data for the most successful airframe firm of the pre-expansion years are available, however, and if this example is taken as a maximum for the principal 15 contractors it is reasonable to conclude that the total area available for production in 1935 was no greater, and probably somewhat less, than two million square feet.[71] By September 1938 the area actually productive amounted to 4.6 million square feet, the average productive area of 22 firms being 209,000 square feet. The largest privately owned factory in the industry was that of Armstrong Whitworth, whose Whitley bomber required a productive area of more than 455,000 square feet.[72]

Despite the existence of excess capacity at the beginning of rearmament, the manufacturing facilities available to the industry proved inadequate even for the comparatively limited requirements of schemes A and C. Handley Page had to extend their factory in August 1935 to fulfil an unexpected contract for the Harrow bomber.[73] At the start of scheme F, Blackburn had four designs accepted by the Air Ministry and the Admiralty, and an agreement was eventually struck with the ship-building company, Denny and Brothers Ltd, under terms of which Denny was to build Blackburn aircraft at a new factory on the Clyde.[74]

At Vickers the calculation was somewhat different. McLean wrote in February 1935 that rearmament orders

will outstrip the capacity of the factory . . . and we shall be faced with a much increased carry-forward into the year 1936/7 . . . Through our not

being able to offer attractive deliveries, the Air Ministry may be impelled
to place work elsewhere, to our detriment now and to our danger in the
future.[75]

Moreover, both the Vickers and Supermarine factories were
badly balanced for the production of large numbers of aircraft.
Supermarine had more assembly capacity than manufacturing
capacity and plans were therefore made to increase manufacturing
floor space by 49 per cent. At Vickers the opposite problem existed,
requiring an expansion of assembly capacity of 45 per cent.[76] Firms
were also influenced by strategic considerations. Air Ministry concern about the vulnerability of southern factories to air attack
helped persuade Fairey to develop new capacity at Stockport. The
management believed that this would 'favourably influence
prospects of future contracts'.[77]

The sums spent by the firms on factory extensions between 1935
and 1938 were very substantial indeed. The total investment in airframe capacity covered by the Capital Clause amounted to £2.6
million by the end of 1936 and to £4.8 million by 1939.[78] Investment
in plant and machinery at both Bristol and Fairey was equivalent
to the total cost of equipping the Austin shadow factory.[79] The
factory space employed by such firms as Boulton Paul and Westland
during rearmament was largely acquired after 1935.[80] Short
Brothers, like Blackburn, associated itself with a shipbuilding firm
and opened a new factory in Belfast.[81]

The process of expansion was not without its difficulties. There
was, for example, some delay before new factories could be fully
equipped with machine tools. The government always recognised
that rearmament would impose a considerable burden on the British
machine tool industry but it tended to underestimate precisely how
great that burden would be. In 1936 the Air Ministry calculated that
the established airframe industry would require machine tools to
the value of £800,000 for scheme F, and an estimate was given of
£210,000 for the two airframe shadow firms.[82] In fact the airframe
firms spent nearly twice this sum, and the value of plant and
machinery required for the shadow factories amounted to
£609,000.[83] Purely quantitative estimates were, in any case, certain
to be misleading. Completely new manufacturing processes had to
be developed by the aircraft constructors because of the changes
taking place in aeronautical technology and this, in turn, involved
the design and construction of new types of machine tools.[84]

In these circumstances it was difficult for the machine tool industry to plan its own expansion. The output of machinery increased by 200 per cent between 1935 and 1939[85] but the manufacturers were understandably reluctant to expand further because of the probability that demand would contract after the completion of the defence programme.[86] They were also pressed by the government to maintain exports, and in response they sent about 30 per cent of their produce abroad between 1935 and 1939. Deficiencies in domestic supply could only be redressed by importing foreign machines.[87]

Nevertheless, there was no general shortage of industrial capacity at this stage of rearmament. The real problem facing the aircraft industry was the absence of technical standardisation, which prevented available capacity from being employed efficiently. Commenting on the scale of the industry's expansion in 1938, the Director General of Production concluded that, with only a few exceptions, 'none of the aircraft firms will be short of the essential floor space required for the execution of the programme'. Nor, despite earlier problems, was there now any shortage of machining capacity.[88]

In general, the largest investment programmes were undertaken by firms which were already well established at the beginning of rearmament. By the end of 1938, 64 per cent of total Capital Clause expenditure had been incurred by Hawker Siddeley, Bristol, Vickers and De Havilland (three of which manufactured both airframes and engines). Past prosperity presumably encouraged confidence in the future. The remaining outlay was divided between 12 companies including those such as Napier, Westland, and Handley Page whose profitability had collapsed during the early 1930s, and those like Airspeed whose principal interest was the civil market.[89] With fewer resources at their disposal, however, these firms were sometimes forced to seek out improvements in efficiency which seemed less important to larger manufacturers.

The extent to which the industry's capacity was underutilised until the later months of 1938 is illustrated by the increasing number of productive workers per square foot (see Table 11) and by the growing efficiency with which floor space was employed. Private investment in the professional airframe industry, combined with state investment in the two airframe shadow factories, had created a total productive area of 4.6 million square feet by September 1938. Total output for 1938 amounted to 9.8 million pounds of airframe structure weight, or to slightly more than 2 pounds per square foot. By September 1939 productive floor space in the airframe sector totalled 8.1 million square feet of which nearly 5 million square feet had been financed by the industry itself. Output had increased to 3.6 pounds per square foot.[90]

TABLE 11

Floor Density (sq. ft. per productive worker) in the
Airframe Industry, 1938–39

	August 1938	December 1939
Floor density for large aircraft	90	85
Floor density for small aircraft	145	115

Source: PRO AVIA 10/267.

New expansion plans, 1938–39

The worsening of international relations, and the commencement of scheme L, forced state and industry to bury their differences and to co-operate more closely in planning expansion. As war became increasingly likely, the industry reconsidered its attitude towards investment. 'If I were a constructor', wrote the SBAC's solicitor, 'I think it very likely that I should not be prepared to incur any further capital expenditure as I should think I had already laid out enough.'[91] The majority of industrialists agreed. One by one they appeared before the Supply Committee between April and June 1938; one by one they declared that they had invested substantial sums in capital extensions, and one by one they refused to invest any more.[92]

It was inevitable that the industry should have attempted to limit expenditure under the Capital Clause as rearmament gathered pace, and the government now declared its willingness to invest directly in buildings and plant for aircraft production. State funding increased the productive floor space of the airframe industry (excluding subcontractors) by 140 per cent between August 1938 and June 1940 and ensured that the expansion of output achieved in 1938 and 1939 was sustained following the outbreak of war.[93]

In looking to the state to finance expansion, the aircraft industry had no choice but to accept a greater measure of Air Ministry control over the way resources were used. For example, the Ministry could now insist on firms building aircraft which were not of their own design, as it did in the case of a new factory provided for

Fairey.[94] The Ministry could also influence the way in which floor space was employed by insisting that new capacity for Gloster, A.V. Roe, and Metropolitan-Vickers was planned for assembly work, while most component manufacture was subcontracted.[95] It could even affect the production methods employed, as with the new Vickers assembly works at Chester which was planned on the understanding that 'progressive' assembly methods would be adopted.[96] Moreover, the introduction of direct state investment provided the most important justification for reducing the profit rates paid to the aircraft industry in 1939.

Until the commencement of scheme L, capital investment and production plans had largely been formulated in isolation. The government distributed contracts and relied on the industry to provide the necessary floor space. From 1938 onwards, however, factory and production planning were closely co-ordinated. The clarification of long-term government requirements and the creation of the Air Council Subcommittee on Supply played a key role in improving relations between the Air Ministry and the firms, and regular contact with the industrialists through the Supply Committee kept senior Air Ministry officials informed about the progress of important projects. This knowledge was then used to correlate investment and production programmes.

For example, the Supply Committee's discussions with Hawker Siddeley and Handley Page soon revealed that the development of the new Manchester and Halifax bombers was likely to be more protracted than originally thought. It became clear that these types would not appear before 1940–41, and alternative plans had therefore to be drawn up to meet the deficiency.[97] The weight of bomber production was at this time to be borne by the proposed Wellington group, but in May 1938 Vickers was still not in a position to release completed drawings to the other firms and, consequently, neither Gloster, Armstrong Whitworth, nor Austin were willing to commit themselves to high levels of Wellington production during 1939.[98] The Air Ministry could only be certain of receiving large numbers of Wellingtons from Vickers in the meantime.

So the Wellington group idea was revised. Gloster was directed to join Hawker in producing the Hurricane.[99] Factory capacity for the new heavy bombers was sanctioned but, until these types could be introduced, Handley Page, Armstrong Whitworth and Austin were instructed to continue producing the Hampden, Whitley and Battle bombers. All Wellington production was now to be undertaken by the armaments manufacturer, Vickers-Armstrong – parent company of the two Vickers aircraft firms. At the end of September Freeman approached Sir Charles Craven and proposed that 'the wide organisation and experience of the Vickers-Armstrong group might profitably be applied to the setting up of a comprehensive scheme designed to accelerate production.'[100]

Craven agreed to co-operate fully. The new organisation was placed under a senior director of the Vickers-Armstrong group and responsibility for production vested in the Vickers Aviation manager, Trevor Westbrook. An aerodrome and erecting shops were built at government expense, the highest possible proportion of component manufacture was subcontracted, and the existing Vickers-Armstrong subsidiaries were absorbed by their parent company for management purposes.[101] Sir Robert McLean was fired and Vickers Aviation and Supermarine were formally wound up.[102]

During 1939 an extensive network of subcontractors was established in the north-west to supply Wellington parts to an assembly factory in Chester. This was augmented by a similar network centred on Blackpool later in 1939 and by increased capacity at Vickers-Armstrong's Weybridge factory. Excluding subcontractors, the scheme involved an increase in the productive floor space for Wellington production of 450 per cent.[103] The Chester and Blackpool factories were originally intended to produce a total of 100 aircraft per month, although the target was raised to 150 during the first year of the war.[104] The plans ensured that the Wellington was available in quantity to Bomber Command until re-equipment with the new heavy types began in 1941.

Production problems overcome

Although the state could play an important part in laying the industrial foundations for wartime aircraft production, the technical problems which delayed the introduction of some of the most important designs during the early stages of rearmament could only be solved by the aircraft industry itself. By 1938, however, many of these difficulties were finally being surmounted, and the firms were embarking on the more straightforward task of producing relatively standardised designs. The Hurricane and the Hampden entered production, the Spitfire and the Wellington appeared, and A.V. Roe and Rootes produced their first Blenheims.

Aircraft production could now be accelerated simply by creating more floor space and by duplicating drawings, jigs, tools and machinery. The English Electric Company sent workmen to Handley Page to manufacture a set of standard parts for the Hampden.[105] Hurricane jigs and tools manufactured by Hawker and A.V. Roe enabled production at Gloster to start just 12 months after orders were placed. Vickers-Armstrong's Weybridge factory and its subcontractors supplied parts and components to the new assembly plant at Chester allowing the first 'shadow' Wellingtons to be assembled during the final months of 1939.[106] The extent to which these preparations accelerated production at Vickers-Armstrong is illustrated in Table 12.

TABLE 12

Wellington Production at Vickers-Armstrong

Months from Start of Production	3	6	9	12
Weybridge Production (Cumulative)	29	86	146	203
Chester Production (Cumulative)	12	99	303	489

Source: PRO AVIA 10/311; PRO AIR 19/524.

TABLE 13

Fixed Capital Expenditure at Bomber Firms, 1935–38

Firm	Expenditure (£)
Bristol (Airframes)	853,000
Fairey	381,000
Vickers	347,000
Armstrong Whitworth	269,000
Handley Page	163,000

Source: PRO CAB 16/227.

TABLE 14

Per capita Sales, Handley Page and Armstrong Whitworth

	Sales, 1937 (£)	Labour, June 1937	Per capita Sales (£)
Armstrong Whitworth	1,148,676	3,546	324
Handley Page	1,311,335	2,176	603

Source: Sales figures from Handley Page and Hawker Siddeley archives, company accounts for 1937; labour figures from PRO AIR 6/53. Armstrong Whitworth sales figures is the average of two financial years to 31 July 1937 and 1938; Armstrong Whitworth sales in the year ending 31 July 1937 were only £729,878.

TABLE 15

Index of Handley Page Hampden Labour Costs

Cumulative Aircraft	Labour Cost per Aircraft
25	100.00
50	82.50
100	71.25
200	68.75
350	65.00

Source: Calculated from cost records in AC 70/10/139 (labour cost equated to 100 at £4000).

TABLE 16

Index of Bristol Production Costs

Cumulative Aircraft	Labour Cost per Aircraft	Complete Aircraft Cost
25	100	100
50	76	75
100	61	62
200	52	58
400	44	50
800	37	44

Source: Connolly Papers, data supplied by H.J. Pollard of the Bristol Aeroplane Company (unspecified cost units, labour cost equated to 100 at 25 cost units, complete aircraft cost equated to 100 at 45 cost units).

Yet the increased output cannot be explained by design stability alone, for production and design methods were also being improved. Some of the earliest breakthroughs came at Handley Page. The 'demobilisational instability' which followed the First World War had brought Handley Page to the verge of collapse, and there had been considerable fluctuations in the company's fortunes since then.[107] The years immediately preceding rearmament were the worst since the early 1920s,[108] and it is hardly surprising that at the beginning of scheme C the decision was taken to avoid investing in buildings and plant which might be surplus to requirements at the end of rearmament (see Table 13).[109]

It was thus particularly important to use available resources as economically as possible. Handley Page's solution was a production technique called 'split' construction, in which the main components of the aircraft were literally split up into assemblies and sub-assemblies. The traditional system of working progressively on large numbers of aircraft over a protracted period was abandoned. Assembly jigs, which determined the rate of production and the floor space required, were now used for assembly alone. Equipment and services were installed into sub-assemblies where they were most accessible to the work-force, allowing final assembly time to be reduced. As many man-hours as possible were concentrated simultaneously on individual aircraft, and the smallest possible number of aircraft were worked on at any given time for a given rate of output. This reduced production times, floor space requirements, and the capital invested in work-in-progress.[110] Quite how effective these methods were may be gauged from the fact that Handley Page achieved a production rate of eight Harrow bombers per month using a single assembly line with just one set of fuselage jigs. In contrast, the production of Armstrong Whitworth's Whitley bomber at a lower rate required an enormous layout consisting of 16 fuselage jigs.[111] Per capita output at Handley Page was nearly double that of Armstrong Whitworth in 1937 (see Table 14).

Weir visited Handley Page's factory in November 1936 and was clearly impressed by what he saw. He wrote:

The Cricklewood factory is now equipped and organised to ensure effective production of large metal stressed skin airframes. It is clear that a definite systematic production technique has been developed . . . and that effective production from the floor space will result . . . The jig system and sub-assembly technique appear so effectively worked out as to give the impression that the problem of producing large machines has been solved.[112]

In 1937 this may have represented an isolated case but it augured well for the future by introducing a method of production which brought demonstrable improvements in efficiency. By the time the Hampden was introduced in 1938, split construction had been developed further, and when the deteriorating international situation led the Air Ministry to place larger contracts the economies of the system were even more clearly demonstrated (see Table 15).[113] Moreover, the absence of work-in-progress also reduced Handley Page's requirements for working capital. The costs incurred by the time Hampden deliveries started amounted to 23 per cent of total contract expenditure.[114] At Vickers, approximately 40 per cent of the total production costs of the initial Wellesley contract had been incurred before deliveries began.[115]

Although Handley Page solved some of the most difficult production problems a year before most of the other main companies, there is evidence that significant progress was being made elsewhere. Bristol recorded continuous reductions in manufacturing costs throughout the rearmament years (see Table 16). After a visit to Supermarine late in 1937, Sir Charles Craven wrote to Vickers' works manager, Trevor Westbrook: 'I wish to congratulate you on the great change I saw, especially as regards the way you have opened up the shops.'[116] Hawker introduced new assembly methods in September 1938 which 'resulted in a definite increase of output'.[117] Short Brothers employed the management consultants, Production Engineering Ltd, whose investigations led to 'a decided increase in output [of the machine department]' and a fall in labour costs.[118] At Armstrong Whitworth the problem of supervising large-scale production was assisted by the introduction of standard costs between 1938 and 1940.[119]

Some of the most significant developments in aircraft design and production came at A.V. Roe and were embodied in a bomber designed to the Air Ministry's specification P13/36, which became the Manchester, and subsequently the Lancaster.

In 1938, A.V. Roe's new assistant chief designer made his first acquaintance with the Manchester prototype. 'Although apparently following conventional lines,' he wrote later, 'the basic structure was outstanding in its suitability for quantity production by semi-skilled labour.'[120]

From the outset, the Manchester had been designed with ease of production in mind. Split construction was employed to improve accessibility to labour, and great emphasis was placed on simplifying the detail design and on reducing the number of different parts. This approach entailed frequent painstaking compromises between the designer's attempts to save weight and the production engineer's desire to reduce man-hours, but its implications were far reaching. In the past, the importance which British firms had attached to saving weight had resulted in the design of complex airframe structures which were expensive to produce without special machinery. But A.V. Roe calculated that, in return for an insignificant increase in weight, the typical production time for a medium bomber could be cut by more than 20 per cent.[121]

New attitudes towards factory layout are also in evidence at A.V. Roe. An article published in 1939 noted that labour had been saved by replacing the 'departmental' outlook with the production outlook, by ensuring a flow of work in sequence from one operative to the next, and by organising final erection and assembly operations so that airframes were kept on the move to a regular time schedule. It was concluded that 'the technique of the conveyor can be used even if that mechanism itself is not at present suitable for aircraft work'.[122]

A.V. Roe gained invaluable experience in quantity production methods in the early 1930s, but there was also a tradition of close co-operation between the company's design and production staff.[123] At other firms these links were sometimes less well established and had to be strengthened during the rearmament years. For example, Westland's organisation proved unsuited to coping with Air Ministry demands after 1935, and in 1939 a new administrative structure was introduced based on that of the American motor company, Chevrolet. The key feature was the creation of a 'production design' department which improved the co-ordination of Westland's design and production work so successfully that the new organisation remained unmodified throughout the war.[124]

Product design, factory layout, and corporate structures all had to be modernised to cope with the rearmament programmes. But rearmament also compelled the aircraft industry to adopt new production processes. For much of the period aircraft production remained characterised by a very high degree of handwork. Skilled sheet metal workers were employed on metal-forming operations which had long since been mechanised in mass-production industries. There were good reasons for this. First, the high-capacity hydraulic power presses required for metal forming were far more expensive than ordinary machine tools.[125] Second, the Air Ministry believed that work of this nature could be subcontracted and that the necessary machinery should not, therefore, rank for consideration under the Capital Clause.[126] Third, the presses had to be used in conjunction with tools which were, in themselves, so expensive that their use could only be justified when long and uninterrupted production runs were in prospect.[127]

By 1938, however, skilled sheet metal workers were in short supply, aircraft were being ordered in larger quantities, and several important designs had been standardised. At the same time, firms such as A.V. Roe and Blackburn had established that hydraulic presses equipped with inexpensive rubber dies could be used in aircraft manufacture. The presses produced remarkable economies and allowed aircraft to be designed with fewer parts.[128] As the potential of press work became clearer, the Air Ministry altered its stance and from August high-capacity hydraulic presses were brought within the terms of the Capital Clause.[129] From then on, the mechanisation of this particular stage of production proceeded steadily.

While the professional aircraft industry was developing more modern production methods, the first airframe shadow factories exploited many of these techniques from the very beginning. In 1936, Rootes had expected that most of the technical and planning information required for their Blenheim factory at Speke would be supplied by Bristol. Bristol, however, was preoccupied with its own production problems, so the Rootes organisation eventually undertook most of the work itself, introducing appropriate adjustments to suit the greater volume of output planned for the shadow works.[130] This factor, coupled with a relative absence of financial restrictions on construction and equipment resulted in the creation of what, for its time, was one of the largest and most modern aircraft factories in the world. Rootes organised a single straight-line production flow from one end of the factory, where materials were received, to the other, where completed aircraft were wheeled out to the aerodrome; assembly methods were used which closely resembled those developed by Handley Page, and the entire organisation was regulated by a particularly advanced system of manufacturing control.[131] In 1939, the two shadow airframe firms delivered no fewer than 970 aircraft, 12 per cent of total output.[132]

The successful application of modern mass-production techniques to aircraft at Austin and Rootes led to the recruitment of other companies from both the motor and electrical industries. Such firms were to play a crucial role in managing the expansion of aircraft production after the outbreak of war.

Subcontracting

The Air Ministry's decision to subcontract airframe production was discussed in the previous chapter. Subcontracting had been a key feature of the Ministry's war production plans since the 1920s, but it could not have succeeded without a degree of design standardisation and co-operation between state and industry which was lacking during the early stages of rearmament. The dangers of attempting to subcontract undeveloped aircraft were vividly illustrated by Supermarine's experience with the Spitfire. In this instance, problems which were usually contained within the main airframe factories were distributed by the Air Ministry among no fewer than ten different subcontractors after it was concluded that Supermarine (who already had contracts for two other models) was incapable of producing 270 Spitfires by the scheme F target date of April 1939.[133]

The Spitfire's fuselage was not particularly complex and did not differ greatly in construction from other fighter aircraft of the period, but the famous elliptical wing sacrificed ease of production to maximise strength, weight-saving, and aerodynamic efficiency and to accommodate the eight-gun armament required by the RAF.[134] The prototype Spitfire flew in March 1936, but prototypes usually differed significantly from production aircraft and it is probable that the design was far from complete when the Air Ministry decided to resort to subcontracting.[135]

Production drawings were prepared under great pressure by a relatively small technical team and were then circulated to the sub-contractors. Supermarine staff frequently visited the other factories and monthly meetings were held between the various Spitfire manufacturers and DAP representatives at Supermarine's works.[136] Yet there was always a danger that design mistakes might slip through and, in the event, an error in the construction of the wings only came to light shortly before the first production aircraft was due to be delivered.[137] Many subcontracted components had to be scrapped. The subcontractors claimed that Supermarine had issued inaccurate drawings and had not kept them informed about modifi-cations. Supermarine, on the other hand, blamed the subcontractors for not being sufficiently accurate in construction.[138] The first Spitfire production programme envisaged that 270 aircraft would be completed by April 1939 whereas only 130 were actually delivered.[139]

The Air Ministry's declared policy prior to rearmament was that aircraft work should not be subcontracted until the accuracy of production drawings had been proved.[140] Subcontractors like Folland Aircraft actually refused to undertake work for one leading company because 'the drawings were not complete and modifica-tions [were expected to] take place causing serious delays'.[141] The difficulties encountered with the Spitfire suggest that Folland's caution was amply justified.

By the spring of 1938, however, the standardisation of most important aircraft designs had eliminated many of the risks involved in subcontracting. The aircraft firms had often opposed sub-contracting in the past but, during the spring of 1938, recession in the general engineering industry led to growing pressure on the principal Air Ministry suppliers to distribute rearmament sub-contracts. The EEF urged the government to adopt subcontracting as official policy, and the SBAC found itself being approached by so many engineering companies that in May 1938 it created a register of subcontracting capacity.[142]

At the same time, several firms realised that labour shortages would prevent them from achieving their 1939 production targets unless the amount of subcontracting was increased.[143] For example, the burgeoning RAF demand for interceptor fighters compelled Hawker to implement a phased expansion of Hurricane sub-contracting in the autumn of 1938. Plans were initiated to raise the proportion of subcontracted man-hours from 25 per cent in 1939 to 50 per cent in 1940.[144]

Thus the amount of subcontracting was already being increased when, in September 1938, the Air Ministry formally requested the industry to subcontract at least 35 per cent of airframe man-hours.[145] Companies like A.V. Roe co-operated readily. The general manager, Roy Dobson, wrote to Lemon that

as soon as our conversations on this matter began . . . we took from our shops a lot of drawings and dispatched them [to the subcontractors]. This has left us short but we are putting up with it and not grousing, because we thought it was all to the good of the cause.[146]

Elsewhere, however, there was reluctance, or even resistance, but not without good reason. Richard Fairey wrote to Bruce-Gardner that

although we and many others are doing our best to reach a solution on these lines, I am rapidly becoming convinced that it will not be successful . . . There is too much pressure on suitable subcontractors and unsuitable ones take too much coaching.

Management could, he suggested, be more productively employed in Fairey's own factory.[147]

At Handley Page, production plans to build the Hampden bomber at the rate agreed with the Air Ministry had been completed by the time the firm was asked to increase the amount of sub-contracting. Frederick Handley Page therefore suggested that the English Electric Company should, instead, build any additional Hampdens required by the RAF and that subcontractors should be employed on jig and tool production for the Hampden and its successor, the Halifax. As the Air Ministry was, in fact, already involved in discussions with English Electric, both Handley Page's proposals were quickly accepted.[148]

Subcontracting was by no means a panacea. There was initially some delay before subcontractors could match the efficiency of parent firms. It was for this reason that the productivity (in terms of output per jig) recorded at six companies building Hurricane wings in 1940 varied by as much as 50 per cent.[149] Moreover, it will be shown in a later chapter that subcontracting imposed severe long-term constraints on wartime aircraft production. The important point, however, is that between 1938 and 1940 subcontracting achieved what it was intended to achieve by providing a quick solution to the immediate problem of labour absorption. By March 1940, 50,500 workers were employed on airframe subcontracts. This amounted to 27 per cent of total airframe employment.[150] In other words, if approximately 10 per cent of airframe production was subcontracted prior to the summer of 1938, the extension of subcontracting thereafter may have increased the output of the industry by as much as 17 per cent by the summer of 1940.

Subcontracting brought other advantages. It resulted in a major extension of credit between parent firms and their suppliers, which became an increasingly important source of finance for aircraft production; and it tapped, at minimal capital cost to the government, managerial and factory resources which might otherwise have been impossible to transfer to the aircraft industry. The cost per square foot of adapting general engineering firms for aircraft production was less than one-sixth of the cost of floor space at new aircraft factories.[151]

BRITISH AIRCRAFT PRODUCTION ON THE
OUTBREAK OF WAR

In 1939, average employment in the airframe industry (including subcontractors) was double the level of 1938 but the total output of the industry increased by nearly 200 per cent during the same period (see Table 4). In the final months of 1938 production overtook the Air Ministry's September programme, and from January to September 1939 the industry delivered a total of 5797 aircraft – 633 more than the official target.[152] A productive floor space of 8 million square feet was in operation, and plans were in progress to increase this figure to 19 million square feet by 1941, initially to augment the output of existing models like the Wellington and subsequently to provide capacity for the new heavy bombers.[153]

Important lessons had been learned. To avoid any repetition of the subcontracting problems encountered with the Spitfire, Disney recommended that 'the first few, say 50, of any new type of aircraft must be built by the designing firm by whom all jigs and tools must be not only designed but tried out'.[154] In 1938 the Air Ministry began to ration experimental projects to avoid over-taxing the industry's design resources.[155] The failure of the De Havilland Don illustrated clearly how excessive service requirements could reduce the quality of aircraft, particularly aircraft of wooden construction, and it seems certain that this experience influenced the later decision to install the bare minimum of equipment into De Havilland's B1/40 (which became the Mosquito).[156]

The aircraft industry was by no means fully prepared for war production in September 1939, for the war was accompanied by many new obstacles which were often impossible to foresee in peacetime. The principal accomplishment of the rearmament years lay in the introduction of aircraft like the Spitfire, the Hurricane, and the Wellington after a difficult and protracted period in which extensive production and investment programmes were frustrated by problems associated with the transition from the biplane to the monoplane. With the finalisation of these designs an increase in production could be obtained simply by creating the necessary additional capacity, like the Castle Bromwich Spitfire factory or the two new Wellington factories in the north-west, or by subcontracting.

Yet important developments in design and production techniques demonstrated that there was as much to be gained from improvements in efficiency as there was from the creation of new capacity. As the supply of labour, machine tools and floor space became more constricted during the war itself, the wider application of these methods played a crucial role in maintaining the expansion of output.

NOTES

1. Weir 19/2, Weir to S of S, 10 June 1935.
2. Calculated from aircraft delivery figures in PRO AIR 19/524. Hawker types were being built by a number of different firms.
3. Weir 19/21, Fairey to AMSO, 24 February 1936.
4. PRO AIR 2/1790, DAP to AMSO, 11 May 1936. Although bombers like the Bristol Blenheim were known as 'medium' types in 1936, they were increasingly re-categorised as 'light' as the RAF specified larger and heavier aircraft.

5. Ibid; PRO AIR 19/524.
6. Ibid.
7. 150 Battles and Blenheims were ordered, 100 Handley Page Harrows, 96 Vickers Wellesleys, and 80 Armstrong Whitworth Whitleys.
8. AC 70/10/55, memorandum by the SBAC, 'The Problem of Efficient Quantity Production of Modern Service Aircraft', 1937.
9. PRO BT 28/423, report entitled 'Reduction in the Time of Change-Over to a New Type of Aircraft', by Sir Ernest Lemon, 12 August 1942. Lemon was at this time industrial adviser to the Ministry of Production. See also PRO AVIA 46/88 for Bristol's comments on the report.
10. EEFAW, MSS 237/1/6, ACM NTC meeting, 30 May 1935.
11. PRO AIR 6/44, Bristol form 1407, 15 February 1936; Fairey form 1407, 8 February 1936.
12. E. Mensforth and W.E.W. Petter, 'Aspects of the Design and Production of Airframes with Particular Reference to their Co-ordination and to the Reduction of the Development Period', *Journal of the Royal Aeronautical Society*, 48 (1944), p. 242.
13. AC 70/10/55, memorandum by the SBAC, 'The Problem of Efficient Quantity Production of Modern Service Aircraft', 1937; PRO AIR 6/58, memorandum by DGP, 12 July 1939.
14. PRO BT 28/423, report by Sir Ernest Lemon, 12 August 1942.
15. PRO AIR 2/1790, DAP to AMSO, 11 May 1936 (two of the types ordered from Blackburn were, however, 80 per cent interchangeable). The Air Ministry had detailed information about drawing office employment in the industry. See Weir 19/23A, Air Ministry statement on employment in the aircraft industry, 30 June 1936.
16. PRO AVIA 46/114, information supplied by Hawker to the official historian in 1943.
17. PRO AIR 6/25, S of S EPM, 21 April 1936; PRO AIR 6/26, S of S EPM, 22 June 1936.
18. AC 70/10/67, Handley Page to Bruce-Gardner, 3 March 1938.
19. AC 70/10/55, Burroughs (Gloster) to SBAC, 16 November 1937.
20. PRO AIR 6/33, S of S EPM, 5 April 1938.
21. PRO AVIA 46/110, Resident Technical Officer to Air Ministry, 9 September 1937.
22. AC 70/10/67, Handley Page to SBAC, 14 June 1938; PRO AIR 6/51, note by AMRD, 3 December 1937.
23. AC 78/23, comparison of design effort prepared by Handley Page's outside production office, 1942.
24. PRO AVIA 46/113, Official Historian's type biography of the Hampden; AC 78/23, papers on the Hampden.
25. AC 70/10/159, Ratcliffe to Handley Page, 11 December 1934.
26. AC 70/10/159, drawing office to managing director, 22 March 1935.
27. AC 70/10/159, design engineer to managing director, 6 February and 13 March 1936; sales manager to managing director, 13 June 1936.
28. AC 70/10/67, production engineer to managing director.
29. AC 78/23, papers on the Hampden.
30. S.D. Davies, 'Aeroplane Design and Production', *Aircraft Engineering*, 11 (1939), p. 121.
31. Vickers 780, J.D. Scott's interview with Barnes Wallis, 1 February 1960; Vickers 701, Scott's interview with T.C.L. Westbrook, 15 October 1959.
32. Weir 19/21, Fairey to AMSO, 24 February 1936.
33. Weir 19/20, Weir to S of S, 10 September 1936; *The Aeroplane*, 3 April 1942. The Wellington was originally scheduled to enter production in the spring of 1938. Pressure from the Air Ministry subsequently led Vickers to promise earlier deliveries but these promises proved completely unrealistic. See PRO AIR 6/24, S of S EPM, 25 February 1936, 5 March 1936. See also PRO AIR 6/47, notes on Wellington and Wellesley production by Vickers, 14 December 1936; for the original Wellington programme see PRO AIR 2/1790.
34. For progress reports on the Blenheim see PRO AIR 6/44, Bristol form 1407s, 28 December 1935 to 1 February 1936, 21 March 1936; PRO AIR 6/45, form 1407, 30 May 1936; PRO AIR 6/46, form 1407, 18 July 1936; PRO AIR 6/47, form 1407, 14 November 1936.
35. Vickers 188, Supermarine quarterly report to March 1938.

36. Tuffen and Tagg, *Hurricane*, p. 48; PRO AVIA 15/639, G.P. Kirk to ADAP1, 30 June 1940. Diving restrictions were imposed on the fabric-winged Hurricane.
37. PRO AIR 6/27, S of S EPM, 29 September, 13 October 1936; PRO AIR 6/30, S of S EPM, 8 June 1937; PRO AIR 6/31, S of S EPM, 28 September 1937; Weir 19/21, notes on a visit to Fairey, Stockport, February 1937.
38. PRO AIR 6/49, Clegg to DAP, 5 June 1937.
39. PRO AIR 6/46, Bristol form 1407, 5 September 1936.
40. Vickers 189, Supermarine quarterly report to June 1938; PRO AIR 6/35, S of S EPM, 22 September 1938.
41. PRO AVIA 46/114, information supplied by Hawker to the official historian in 1943.
42. PRO AIR 6/30, S of S EPM, 8 June 1937; PRO AIR 6/31, S of S EPM, 27 July, 28 September 1937.
43. PRO AIR 2/1790, DAP to AMSO, 11 May 1936, 27 August 1937; PRO AIR 6/31, note by Weir of an interview with Mr Verdon Smith of the Bristol Aeroplane Company, Appendix C, S of S EPM, 28 September 1937.
44. Vickers 678, J.D. Scott's interviews with McLean, 8 and 9 November 1959.
45. Weir 19/21, Weir to S of S, 17 February 1937; Vickers 687, McLean to Bird, 19 March 1937; Vickers 1371, board meeting, 8 March 1937; Vickers 326, Craven to McLean, 10 March 1937. H.B. Pratt became Supermarine general manager while the former works manager, T.C.L. Westbrook, joined Vickers Aviation.
46. PRO AIR 6/28, S of S EPM, 2 February 1937.
47. PRO AIR 6/30, S of S EPM, 13 July 1937.
48. PRO AIR 6/31, S of S EPM, 28 September 1937, 12 October 1937.
49. D.E.H. Edgerton, 'Technical Innovation, Industrial Capacity and Efficiency: Public Ownership and the British Military Aircraft Industry, 1935–48', *Business History*, 26 (1984), pp. 250–51; Weir 19/13, memorandum by Harold Balfour, 6 July 1935.
50. RRA, Hives file, 'Policy', memorandum by Hives, 4 April 1938.
51. M. Gilbert, *Winston S. Churchill, V, 1922–1939* (London, 1976), p. 915; R.J. Overy, 'The German Pre-war Aircraft Production Plans: November 1936–April 1939', *English Historical Review*, 90 (1975), p. 795. Imperial Airways was also partly state owned.
52. AC 70/10/55, Bowyer to Handley Page, 8 September 1937.
53. PRO AIR 2/2711, DAP to AMSO, 24 May 1938; PRO AVIA 10/217, ACCS meetings, 5 and 6 October 1938.
54. AC 70/10/55, 'The Problem of Efficient Quantity Production of Modern Service Aircraft'; PRO AIR 2/2711, DAP to AMSO, 24 May 1938; PRO AVIA 15/1465, statement by A.V. Roe and Co. Ltd, on a suggested scheme for the formation of central control for the Lancaster group, 5 October 1941; Postan, *War Production*, p. 419.
55. AC 70/10/54, Handley Page to Allen, 28 May 1938.
56. PRO AVIA 10/253, meeting between Deputy Secretary and Sir Frank Spriggs, 18 March 1941.
57. Edgerton, 'Public Ownership', pp. 252–5.
58. AC 70/10/55, memorandum by the SBAC, 'The Problem of Efficient Quantity Production of Modern Service Aircraft', 1937; Cross, *Swinton*, pp. 205–6.
59. PRO AIR 6/51, note by AMRD, 3 December 1937.
60. For the minutes of these meetings see AC 70/10/67.
61. AC 70/10/67, Fryer (Private Secretary to Chamberlain) to Handley Page, 21 December 1937.
62. AC 70/10/67, Handley Page to Chamberlain, 23 December 1937.
63. Weir 19/13, Weir to S of S, 13 December 1937.
64. Weir 19/13, memorandum by Bruce-Gardner, 14 March 1938; PRO AVIA 46/93, Bruce-Gardner to Inskip, 2 April 1938; PRO AIR 6/53, memorandum by the SBAC, 20 April 1938.
65. AC 70/10/54, notes of a meeting between the SBAC and the Air Ministry, 17 January 1938.
66. 'Structure weight' is the net weight of the airframe and excludes ancillary items like engines, propellers, armament, load etc. The published figures also exclude the production of spare parts.
67. Postan, *War Production*, p. 328.
68. PRO AIR 2/1790; PRO AIR 19/524; *Statistical Digest*, pp. 152–3.

69. HSA, Blackburn board minutes, 16 January 1936.
70. Vickers 1392, note on factory extensions by McLean, 11 February 1935.
71. See Hornby, *Factories*, p. 237. The productive area at Hawker between 1934 and 1936 was 146,010 square feet. Data on factory costs between 1935 and 1938 and on floor density in 1938 suggested a productive area for the main 15 firms of between 1.6 and two million square feet in 1935. 'Productive' floor space includes only floor space on which manufacture and assembly is undertaken. It excludes storage, administrative and all other areas not used for production.
72. PRO AVIA 10/267, table by DD Stats. P, 20 December 1941. Additional floor space at Armstrong Whitworth was used for civil aircraft.
73. AC 70/10/159, managing director to works manager, design engineer and production engineer, 1 August 1935.
74. PRO AIR 6/27, S of S EPM, 23 September 1936.
75. Vickers 1392, note on factory extensions by McLean, 11 February 1935.
76. Ibid.
77. *The Statist*, 30 March 1935.
78. PRO T 161/1323/S.40700/42, statement by the Air Ministry of Capital Clause commitments, 28 February 1938; PRO CAB 16/227, statement showing capital extensions expected to rank for consideration under the provisions of the Capital Clause (undated but approximately December 1938).
79. PRO AIR 6/58, memorandum by DGP, 12 July 1939; PRO CAB 102/274, papers prepared for Hornby, *Factories*. Austin: £258,000; Bristol (airframe): £288,000; Fairey: £268,000.
80. *The Statist*, 13 April 1935, 27 July 1935; PRO T 161/1323/S.40700/42, statement by the Air Ministry of Capital Clause commitments, 28 February 1938.
81. PRO AIR 6/45, meeting between Short Brothers and the Air Ministry, 26 June 1936.
82. PRO SUPP 3/9, statement by the Air Ministry of estimated value of machine tools to meet expansion programme, 10 June 1936.
83. PRO T 161/1316/S.40700/7-63, data collected from Treasury Inter-Services Committee files, 9 September 1936 to 29 September 1938; PRO CAB 102/274; PRO AIR 6/58, memorandum by DGP, 12 July 1939.
84. AC 70/10/55, memorandum by the SBAC, 'The Problem of Efficient Quantity Production of Modern Service Aircraft', 1937.
85. Hornby, *Factories*, p. 333.
86. Alfred Herbert Ltd. archive, Acc. 586/30/1, ordinary general meetings, 10 June 1936, 9 June 1937; Federation of British Industries (FBI) archive, MSS 200/F/3/52/15/16, Butterworth Machine Tool Company and Alfred Herbert Limited to the Machine Tool Trades Association, 2 February 1939.
87. PRO CAB 102/511, 'Machine Tools and Rearmament' (unpublished official narrative), section 27. The value of total British machine tool output between 1935 and 1939 amounted to approximately £51 million; the value of exports was more than £15 million. The value of machine tool imports increased from £1.7 million in 1935 to £7.7 million in 1939.
88. PRO AIR 6/54, memorandum by DGP, 5 September 1938; PRO AIR 6/35, S of S EPM, 14 September 1938.
89. PRO CAB 102/274; PRO CAB 16/227.
90. PRO AVIA 10/267, statement by the Air Ministry of productive floor area, September 1939; table by DD Stats. P, 20 December 1941; *Statistical Digest*, p. 153.
91. AC 70/10/54, Allen to Handley Page, 26 May 1938.
92. See, for example, PRO AVIA 10/151, ACCS meeting, 29 April 1938, statements by Spriggs and McLean.
93. PRO AVIA 10/267, table by DD Stats. P, 20 December 1941.
94. PRO AVIA 10/151, ACCS meeting, 2 May 1938; PRO AVIA 10/154, ACCS meeting, 5 August 1938.
95. PRO AVIA 10/154, ACCS meeting, 29 July 1938.
96. PRO AVIA 10/217, ACCS meeting, 1 December 1938. Progressive assembly involved the movement of aircraft subassemblies through successive stages of final assembly and erection.
97. PRO AVIA 10/151, ACCS meeting, 29 April 1938.

98. PRO AIR 2/3273, notes of a meeting between DAP and five aircraft firms, 7 April 1938.
99. PRO AVIA 10/155, 2nd DUS to Hawker Siddeley, 6 October 1938; PRO AVIA 10/9, Spriggs to AMDP, 26 September 1938.
100. PRO AVIA 10/217, ACCS meeting, 5 October 1938.
101. PRO AVIA 10/217, ACCS meeting with Craven, 6 October 1938.
102. Vickers 678, J.D. Scott's notes on McLean's career at Vickers.
103. Calculated from PRO AVIA 10/267, table by DD Stats. P, 20 December 1941.
104. PRO AVIA 10/217, Dunbar (Vickers) to Riverdale, 16 April 1940; MAP to Vickers-Armstrong, 16 July 1940.
105. AC 78/23, papers on the Hampden and Halifax group.
106. PRO AVIA 10/9, Spriggs to AMDP, 26 September 1938; Vickers 196, Vickers-Armstrong Chester quarterly report to March 1940.
107. Fearon, 'Handley page', pp. 66–78.
108. Ibid., p. 72.
109. AC 70/10/54, Handley Page to Allen, 2 November 1935.
110. AC 70/10/67, Handley Page to Bruce-Gardner, 22 February 1938; D.C. Robinson (works superintendent at Handley Page), 'Some Developments in Aircraft Production', *Journal of the Royal Aeronautical Society*, 53 (1949), pp. 39–66.
111. Weir 19/1, DAP to Weir, 2 July 1937.
112. PRO AIR 6/47, notes on a visit to Handley Page Ltd. by Weir, 26 November 1936.
113. Robinson, 'Aircraft Production', p. 44.
114. Calculated from Handley Page works order cost records, Handley Page archive (no accession number).
115. Vickers 324, order books.
116. Vickers 626, Craven to Westbrook, 13 October 1937.
117. HSA, Hawker board minutes, 5 September 1938.
118. RRA, Hives file 'Production Engineering Ltd', Gouge (Short Brothers) to Hives, 16 September 1937.
119. PRO AIR 20/2379, Sir Harold Howitt (financial adviser to the Air Council) to 1st DUS, 2 February 1940.
120. Letter by S.D. Davies (former assistant chief designer at A.V. Roe) to the author, 8 May 1990; 'Personal history of S.D. Davies at A.V. Roe (Manchester), 1938–44', autobiographical account of S.D. Davies' work at A.V. Roe, sent to the author in June 1990.
121. Davies, 'Aeroplane Design', pp. 121-4; interviews with S.D. Davies, 21 and 30 April 1990.
122. Davis, 'Aeroplane Design', p. 122.
123. PRO AVIA 46/105, Postan's interview with Roy Chadwick, chief designer at A.V. Roe, 6 January 1944.
124. Mensforth and Petter, 'Design and Production', pp. 218–20.
125. A comparatively small 2500 ton press cost £18,500. See PRO AVIA 10/183, ASB meeting, 7 May 1941.
126. PRO T 161/1325/S.40700/56, Air Ministry memorandum for the Treasury Inter-Services Committee, 3 August 1938.
127. PRO AVIA 10/189, memorandum by DGAP for ASB meeting, 11 August 1941.
128. PRO AVIA 10/183, memorandum by DGAP for ASB meeting, 7 May 1941; PRO AVIA 10/186, memorandum by DGAP for ASB meeting, 30 September 1941; PRO AVIA 10/189, memorandum by DGAP for ASB meeting, 28 November 1941; Davies, 'Aeroplane Design', p. 124.
129. PRO T 161/1325/S.40700/56, Treasury Inter-Services Committee meeting, 11 August 1938.
130. Weir 19/21, Rootes to AMSO, 27 April 1937.
131. 'The Speke Shadow Factory', *Aircraft Production*, May 1939, p. 230; 'Producing the Shadow Blenheim', *Aircraft Production*, May 1941, p. 10; 'The Control of Planned Production', *Aircraft Production*, April 1941, pp. 120–22.
132. PRO AIR 19/524.
133. Vickers 687, J.D. Scott's interviews with McLean, 8 and 9 November 1959; Vickers 746, memorandum by H.E. Scrope (company archivist), 21 August 1957; PRO AIR 2/1790, DAP to AMSO, 27 August 1937.
134. PRO AVIA 15/2660, Report by North American Aviation on weight comparison

between Mustang and Spitfire, February 1943.

135. E. Mensforth, 'Airframe Production', *Proceedings of the Institution of Mechanical Engineers*, Vol. 156, No. 1 (1947), p. 27. According to Mensforth, 180,000 man-hours were typically required to design a single-engined fighter. The Spitfire 1 absorbed 340,000.

136. Vickers 185, Supermarine quarterly report to June 1937.

137. Vickers 866, McLean to Craven, 29 November 1937.

138. Ibid; Vickers 746, memorandum by H.E. Scrope, 21 August 1957.

139. PRO AIR 2/1790; PRO AIR 19/524.

140. PRO AIR 2/714, DDTD to D of C, 20 November 1934.

141. HSA, Folland Aircraft board minutes, 16 June 1938.

142. PRO AVIA 46/93, Ramsay (EEF) to Bruce-Gardner, 14 June 1938; PRO CAB 21/703, meeting between Sir Thomas Inskip and EEF management board, 5 April 1938; AC 70/10/67, Bruce-Gardner to Handley Page, 19 May 1938.

143. PRO AVIA 46/93, Bruce-Gardner to Kingsley Wood, 28 May 1938; Bruce-Gardner to Ramsay, 15 June 1938.

144. HSA, Hawker board minutes, 19 September 1938, 3 October 1938.

145. PRO AVIA 46/93, PUS to Bruce-Gardner, 13 September 1938.

146. PRO AVIA 10/155, Dobson to DGP, 27 September 1938.

147. AC 70/10/67, Fairey to Bruce-Gardner, 27 October 1938.

148. AC 70/10/79, Handley Page to DGP, 21 and 29 October 1938, 12 December 1938; DGP to Handley Page, 14 December 1938; PRO AVIA 10/151, ACCS meeting, 9 June 1938.

149. PRO AVIA 15/639, meeting held at Gloster Aircraft, 27 June 1940.

150. PRO CAB 102/275, table by D Stats. P, 9 April 1940.

151. PRO AIR 6/58, memorandum by DGP, 12 July 1939.

152. Postan, *War Production*, pp. 472, 484.

153. PRO AIR 6/58, memorandum by DGP, 12 July 1939.

154. PRO AIR 2/2711, DAP to AMSO, 24 May 1938.

155. Postan, Hay and Scott, *Design and Development*, p. 38.

156. PRO AIR 6/33, S of S EPM, 5 May 1938; PRO AVIA 46/116, Official Historian's type biography of the Mosquito.

4

Aero-engine production, 1935–39

Sir Alec Cairncross has recently written that the aero-engine

lies at the heart of aircraft design, production and programming. This is partly because of the long interval between the inception of the design of a new engine and its subsequent production in quantity. This makes the aero-engine the most difficult part of an aeroplane to vary quickly in supply and compels an effort to adjust the programme for the aircraft to the engines available or in prospect instead of the other way round.[1]

This view was certainly applicable to British aircraft production during the Second World War but in the rearmament years air-frames initially presented much greater design and development problems than aero-engines. The central position of the aero-engine in a balanced aircraft production programme was only recognised after the acceleration of rearmament in 1938.

As described in Chapter 1, aero-engine manufacturing in inter-war Britain was undertaken by five companies: Rolls-Royce, Bristol, Napier, Armstrong Siddeley and De Havilland. By the mid-1930s, however, Rolls-Royce and Bristol dominated the military market, and between April 1936 and May 1939 their engines accounted for 83 per cent of total British output in terms of horse-power.[2] The history of aero-engine production in the rearmament years is thus very largely the history of Rolls-Royce, which specialised in liquid-cooled engines, and of Bristol, which manu-factured air-cooled designs.[3]

After the outbreak of war in 1939, Rolls-Royce engines played a crucial role in maintaining the quality of British first-line aircraft, but in quantitative terms demand was relatively modest at the beginning of rearmament. One specific objective of scheme F was to replace the Kestrel-engined Hawker biplanes which had formed the mainstay of the RAF first line during the early 1930s. The new Merlin engine was required for the Battle bomber and for the Hurricane and Spitfire fighters, but all three were single-engined models and the Air Ministry believed that peacetime requirements could be fulfilled without any special measures to expand capacity. Large-scale expansion would in any case have been problematic, for the Merlin had not yet been standardised for manufacture in 1936. Production therefore remained firmly under Rolls-Royce's own control.

In contrast, Bristol's more standardised Mercury and Pegasus designs were required urgently and in much larger quantities for a variety of different aircraft. By far the largest demand came from the twin-engined Blenheim bomber, which was ordered from Bristol and Rootes in 1935 and 1936 respectively, and from A.V. Roe in 1937. These engines, however, also powered the Wellesley, Wellington I, Harrow and Hampden bombers, Gloster's Gauntlet and Gladiator fighters, Fairey's Swordfish torpedo bomber, and the Short, Supermarine and Saunders flying boats. As Bristol lacked the resources to meet the Air Ministry's demands by April 1939, it was decided to activate the Committee of Imperial Defence's plans for using sources of supply from beyond the professional aircraft industry and so the first aero-engine shadow scheme was launched.

This was a potentially dangerous situation. First, as already noted, the rate of obsolescence was particularly high in the mid-1930s and the engines then in production were nearing the end of their development. Second, new aircraft designs increasingly incorporated liquid-cooled engines like the Merlin rather than Bristol's air-cooled models. Liquid-cooled engines were better suited to the specifications issued by the Air Ministry in 1936 and 1937; they had superior aerodynamics because their frontal area was smaller than that of equivalent air-cooled engines and they offered a more direct path to improved aircraft performance at higher altitudes.[4] Moreover, it soon became clear that the Merlin could be used to improve the performance of otherwise obsolete aircraft. The service life of Armstrong Whitworth's Whitley bomber (originally designed around an Armstrong Siddeley power plant) was significantly extended by the installation of the Merlin, and Air Ministry planning in 1939 was based on the assumption that the Merlin would power later versions of the Wellington.[5]

This is not to say that the first shadow venture was worthless. On the contrary, both Bristol and the motor firms gained valuable experience which was put to good use when a second shadow scheme was established for the more advanced Hercules engine later on, but the urgent short-term requirement for Bristol engines did tend to divert official attention from the burgeoning long-term demand for Rolls-Royce models. It was clear to the Air Ministry that more Rolls-Royce capacity would be needed under wartime conditions, but the company's resources appeared to be sufficient for the peacetime programme. Even after the *Anschluss*, both Lord Weir and the AMSO considered that 'so far as could be seen, the existing Rolls-Royce factory should be just about sufficient for all engines of Rolls-Royce types which would be required under the existing and any future programme at present contemplated'.[6]

The issue was only decided by the introduction of scheme L, with its increased provision for Merlin-engined fighters. A new factory was immediately authorised at Crewe, and in 1939 'war potential' planning, not least for new four-engined bombers like the Halifax, led the government to sanction two additional plants.

Nevertheless, Rolls-Royce's productive capacity was far greater in 1938 than it had been in 1935, for, during this period, the company implemented a far-reaching programme of internal reorganisation involving extensive capital investment and the introduction of new management practices. The first section of this chapter examines these issues in detail before considering the development of the shadow scheme and the more ambitious aero-engine production plans inaugurated in 1938–39.

ROLLS-ROYCE AND AERO-ENGINE PRODUCTION, 1935–37

During the first decade of the inter-war period it could not have been predicted that Rolls-Royce would ever become Britain's leading aero-engine manufacturer. The dominant position in the British military aero-engine market was at this time occupied by Napier, and Rolls-Royce's market share actually declined from 25 per cent in 1925 to just 11 per cent in 1929.[7] The subsequent collapse of Napier's fortunes illustrated both the inherent dangers of the military market and the importance of an aggressive approach to research and development. Such an approach produced Rolls-Royce's Kestrel engine at the end of the 1920s and the 'R' engine developed for the Schneider Trophy races of 1929 and 1931. In 1932, work began on the PV12. This was to become the Merlin.[8]

During the early 1930s Rolls-Royce's position was strengthened by growing Air Ministry demands for the Kestrel, which powered aircraft like the Hawker Fury and the Hart. Production rose steadily from 122 engines in 1930 to 1182 in 1935.[9] The service history of the Kestrel reflected the great emphasis placed by British firms on the step-by-step development of established designs, a practice which became particularly important to the improvement of aircraft performance during the Second World War. Between 1928 and 1938 Rolls-Royce produced 4778 Kestrel engines divided between 27 different Marks.[10]

Having increased output significantly between 1930 and 1933, Rolls-Royce had little difficulty in meeting the requirements of the first expansion programme, but a government analysis of war potential undertaken in 1934 concluded that, in wartime, capacity for the Kestrel would be inadequate. It proposed an 'educational' shadow scheme involving the motor manufacturer, Humber. Rolls-Royce was, however, unwilling to disclose design and manufacturing information to a potential competitor and, as the government was not prepared to invest directly in the professional aircraft industry, the company decided instead to increase the capacity of its Derby factory and to subcontract more work. Between 1934 and 1936 Rolls-Royce recruited 1300 extra workers and spent £411,000 on capital extensions. Subcontract man-hours increased from 70,000 in 1934 to 670,000 in 1935.[11]

Under scheme C, orders were placed for 2427 engines, including 190 Merlins.[12] An average production rate of 33 engines per week was envisaged, but as sufficient capacity for this already existed Rolls-Royce's managing director, Sir Arthur Sidgreaves, pointed out to the Air Ministry that the company was in a position to accelerate the completion of its contracts if continuity of orders could be guaranteed after March 1937.[13] Sidgreaves was clearly hoping for additional orders for the Kestrel but, by the time scheme F was inaugurated in the following year, the RAF was more interested in the Merlin, and 2854 of these engines were on order by August 1936.[14]

The full order book and record turnover seemed impressive to contemporary business analysts[15] but the problems which Rolls-Royce had encountered during the 1920s left a lasting impression on the internal administration of the company and when Ernest Hives was appointed general works manager at the end of 1936 he rapidly became convinced that a drastic reorganisation was necessary. Hives had joined Rolls-Royce in 1908. In 1916 he became head of the experimental department, and he remained there while the size of the department and scope of its operations expanded during the inter-war years.[16] However, Hives also had very definite ideas about production. In 1930, while visiting the United States, he had been deeply impressed by American motor manufacturing methods which, with their great emphasis on cost reduction, compared very favourably with British practice.[17] It was no doubt with these comparisons in mind that in January 1937 Hives submitted a report to the Rolls-Royce board which placed the record sales figures in a very different light.

The prosperity of Rolls-Royce's aero-engine division, Hives argued, was based entirely on military sales for the rearmament programme. The company was not sufficiently competitive to exploit the civil market and its fortunes were therefore certain to be reversed after the expansion programme had been completed. Reforms were badly needed in every sphere of Rolls-Royce's activities; its organisational structure made insufficient distinction between motor car and aero-engine operations; not enough attention had been paid to production problems; there was a distinct lack of outstanding personalities among the works management; and supervisors and foremen had all been promoted from the shop floor and were, for the most part, ignorant of modern factory methods.[18]

The lack of attention given to production questions was reflected in the company's works practices. To begin with, the level of capital investment undertaken during the inter-war years had been totally inadequate. In October 1936, 50 per cent of the machinery in Rolls-Royce's factory was over 20 years old. This no doubt helps to explain why the company employed an excessive number of handworkers. The ratio of handworkers to machinists was almost one to one, whereas the modern factory practice was to employ six to eight machine men to every handworker. Such investment as had been undertaken had not always been properly directed. The factory's equipment-ordering department was not in any way responsible for production efficiency and was unfamiliar with works problems. In general, bad factory planning had hindered the flow of work through the shops and the importance of efficient operation planning and liaison between the operation-planning and design staff had gone unrecognised.[19]

Some of these deficiencies might have been remedied had better means existed for the management to monitor the progress of the company's operations, yet in this respect, too, arrangements were far from satisfactory. The rate-fixing department was unable to provide management with an early indication of production costs;

detailed part costs were never available in time to allow production efficiency to be analysed; and the firm still used an antiquated premium-bonus wage system, whereas Hives, rightly or wrongly, favoured the piece-work system which had been adopted by much of the engineering industry by the mid-1930s.[20]

It is possible that Hives exaggerated the extent of Rolls-Royce's troubles, for at the time he assumed his new responsibilities production matters were already causing serious concern.[21] No doubt he was anxious for the Rolls-Royce board to recognise that he had inherited many of the problems likely to afflict the company in his first year as works manager.[22] All the evidence, however, suggests that Hives' critique was, if anything, less than comprehensive. Enormous difficulties were encountered in changing from the standardised Kestrel to the new Merlin. The production manager, Swift, argued that some of this disruption was inevitable but it had been exacerbated by the fact that

Kestrel output was accelerated during 1936 beyond the programme . . . to clear the way for the Merlin [but] the Merlin did not materialise from the production point of view as anticipated, due to unfortunate delays in the design, experimental, and development stages.[23]

The various design and development problems repeatedly held up the completion of the Merlin 1 drawings, and a series of major modifications set production back by several months. Similar delays occurred with the issue of the Merlin 2 drawings between September 1936 and March 1937, and progress was also disrupted by piston modifications. Generally speaking, the problems were less serious from the technical angle but more so from the production viewpoint because the Merlin 2 had been ordered in much larger numbers than the Mark I. Swift was still hopeful that 949 Merlins might be produced in 1937 but he stressed that 'there should be no further serious modification if we are to . . . maintain the estimated output'.[24]

In addition to these internal production problems, the subcontracting network on which the firm had relied so heavily since 1934 was becoming increasingly difficult to maintain. 'There is now intensive competition to get suitable subcontracting firms', Hives wrote. 'Our policy of controlling our own subcontract firms is one of continuous anxiety to me at the present time . . . When the whole of the aircraft industry is competing to get work done outside, it needs constant attention.'[25]

Merlin production had been planned on the basis that 50 per cent of machine hours would be subcontracted, but the reliability of these arrangements was soon called into question. Hives complained that

[w]e have no hold over the firms who take on the work, and we have had several cases where after we have budgeted for certain firms to do certain parts they have failed completely and we have had to take the work away and rush the production, either through our own factory or through some other sub-contractor.[26]

There were particular problems with firms in depressed areas which, Hives believed, had lost the art of rapid production. Of twelve Lancashire companies employed as subcontractors in 1936, all had been abandoned by the end of 1938.[27]

Finally, there were obstacles of a more technical nature. The light alloy castings required for the Merlin's crankcase were produced by highly skilled foundry techniques unsuited to quantity production. Rolls-Royce did not at first consider that the comparatively small total number of engines on order justified any radical modernisation of foundry methods, yet castings production proved the most serious obstacle to the expansion of Merlin output. Rolls-Royce scrapped 65 per cent of the cylinder head castings manufactured for the Merlin 1 and 80 per cent of the first 300 cylinder castings for the Merlin 2.[28]

The result of these numerous complications was a sharp reduction in output during the first half of 1937. Rolls-Royce aero-engine turnover in the year to May was only half that of the year to May 1936,[29] and 1937 was the first year since 1932 in which the size of the work-force remained static.[30] In June, Hives described the Merlin situation as a 'nightmare' but in the same month Rolls-Royce's delivery programme for the year was cut from 827 Merlins to 690 following hold-ups in the airframe sector which were partly the result of the change from the Merlin 1 to the Merlin 2.[31] Extensive night-shifting and overtime brought this target within reach by December.

THE HIVES REFORMS

By January 1938, Rolls-Royce had caught up with the revised programme and steps were being taken to reduce output. Hives wrote:

[f]or six or seven weeks before Christmas we were producing an average of 40 Merlins per week. Our programme is now 25 engines per week. Because of the reduction we have been able to cut out all overtime: we have actually stopped the night shift in some departments and in other departments . . . they are working short time.[32]

By May 1938, Rolls-Royce had delivered over 1000 Merlin engines of which just 160 had flown.[33]

As with the airframe firms, the rapid expansion of output at Rolls-Royce was made possible by the stabilisation of design. Simultaneously, however, a comprehensive programme of internal reform was instigated by Hives with the aim of eliminating many of the deficiencies outlined in his memorandum of January 1937. Hives saw that rearmament provided the opportunity to place Rolls-Royce on a much more competitive footing and to ensure that the company was better prepared to maintain its position in the aero-engine market when the expansion programme came to an end. A major investment programme was launched, albeit under the provisions of the Capital Clause; between December 1936 and May 1938 capital expenditure of £380,000 was authorised by the Rolls-Royce board.[34] Hives commented that

[a] very strict control is now operating as regards ordering new equipment. The supervisor has to state what increase in efficiency we expect to get from a new machine, and after the machine has been working in the factory for two months the results are checked up to see that we are getting the saving we expected.[35]

Meanwhile, the administrative structure of the firm was re-organised in order to segregate the motor car and aero-engine divisions, and new appointments were made accordingly; the most promising supervisory staff were selected to study modern factory methods in Germany and the USA; and the aero-engine factory was completely replanned for Merlin production.[36] 'The essential part of the rearrangement of the shops is to reduce the cost of supervision', wrote Hives, and, when it became clear that a surplus of foremen and supervisors existed, staffing levels were cut.[37]

Operation planning and rate-fixing departments were also re-organised, a new system of cost accounting was adopted, and mechanical accounting equipment was introduced. In October 1937, a firm of consultants, Production Engineering Ltd, was employed to advise on office systems and factory control. 'I am looking forward to the time when I shall know the factory efficiency on Tuesday or Wednesday of the following week', Hives remarked.

'I want the various chiefs to become "cost-conscious" and to realise that no matter how good they think their job is, it is a complete failure unless it is making money.'[38]

During the same period, Hives expanded the activities of Rolls-Royce's subcontracting department. The progress of subcontracted work was carefully monitored and where, as was often the case, orders had been divided between two different firms, their respective delivery records were taken into account when the apportionment of further contracts was under consideration.[39] When, in 1938, the Air Ministry agreed to finance the further expansion of Rolls-Royce capacity, the firm insisted on having the freedom to allocate new machine tools to subcontractors where there was evidence of inadequate machining facilities. This helped to spread labour demands and to reduce Rolls-Royce's vulnerability to German bombing, but it was also remarkably economical in terms of overall capital outlay. Rolls-Royce calculated that 'for every one machine we have loaned to subcontractors we have over 10 existing machines [at the subcontracting firms] working on our pieces'.[40]

Finally, concerted efforts were directed towards the improvement of foundry techniques. The foundry organisation was rationalised and a special section was set up to deal with die-casting. A separate development department was established, together with a new foundry laboratory, and metal patterns were designed for the various parts of the Merlin engine. New inspection methods, gauges and templates were introduced which helped to reduce scrap levels to 12 per cent by October 1937.[41]

Shortly after his appointment, Hives warned the Rolls-Royce board that some time would elapse before his reforms achieved any tangible improvement, and that meanwhile there would be a lot of additional expense.[42] At about the same time he recorded that the company was 'not as well equipped technically or for aero-engine production' as its principal competitor, Bristol. By January 1939, however, he was of the opinion that Rolls-Royce had an aero-engine production organisation more powerful than any in the world 'except Germany'.

The company was ahead of the Air Ministry delivery programme; production and employment had reached record levels, and the administrative changes were generally 'showing excellent results'.[43] The prospect of further expansion, although daunting, could be accepted with greater confidence.

THE BRISTOL SHADOW AERO-ENGINE ORGANISATION

Given the immediate security of government contracts, Rolls-Royce might conceivably have rejected the idea of internal reform in 1937, although at considerable cost in long-term competitiveness. At Bristol the situation was very different. During the 1920s the company had become one of the world's most successful civil aero-engine manufacturers, and its annual civil sales of more than 300 engines per year in the mid-1930s were viewed with envy by Rolls-Royce, which depended entirely on military demand.[44]

An important consequence of Bristol's activities in the civil sphere was that the company paid much more attention to production costs than Rolls-Royce, at least until the advent of the Hives reforms. Yet Bristol was also a leading military manufacturer, and Germany's withdrawal from the Disarmament Conference in 1933 led to a major reorganisation of the company's aero-engine design and production departments.[45] Employment doubled in the following two years, and expansion continued throughout the rearmament period.[46] At the same time an enormous technical effort was devoted to the development of new sleeve-valve engines like the Perseus, the Taurus and the Hercules.[47] As the Air Ministry programme gathered pace, however, the shadow factories became the largest source of supply for Bristol aero-engines.

When the Air Ministry first proposed the creation of a shadow aero-engine industry, Bristol insisted that the motor firms should not be allowed to use the scheme to enter the aero-engine business permanently.[48] Some motor companies were clearly interested in manufacturing complete aero-engines but were deterred by the government's April 1939 deadline for 4000 engines and by the fact that the Air Ministry could only commit itself to ordering 1500 engines of Bristol's Mercury design.[49]

Under these circumstances the shadow firms recommended 'that each of the factories . . . shall undertake the manufacture of a different group of parts of aero-engines and the assembly and test of the complete units shall be undertaken by two of the factories only'. The Air Ministry's preferred alternative, whereby each firm would have made one of two sections of the engine, would 'double the cost of the jigs and fixtures, gauges and tools, add considerably to the amount of plant to be purchased, and might cause delay'.[50] Moreover, such a scheme would have committed Bristol to much more extensive technical supervision, a crucial consideration given the prevailing shortage of higher technical staff. Each different section of manufacture was therefore divided between the participating companies on a man-hour basis (see Table 17).

TABLE 17

The First Shadow Aero-engine Scheme

Firm	Engine Part	Hours
Austin	crankshaft, reduction gear, engine assembly, test	676
Standard	cylinders	676
Rover	connecting rods, pistons, valves, cam	715
Humber	blower and rear cover, petrol pump	700
Daimler	crankcase, oil sump, air intake, rocker gear	729
Bristol	engine assembly/test	250

Source: PRO AIR 2/2577.

Although Bristol forced the Air Ministry to compromise over the organisation of the shadow scheme, the company was in other respects much more co-operative. 'We had a 3½ hours' discussion with Bristols on Friday', Swinton recorded in March 1936.

They brought a complete team who could deal with everything, finance, works management and design. We were able to agree to the general lines of a vast programme involving the creation of a 'shadow industry' by them ... The Bristol plan is much the biggest thing in the whole programme, all through practical and constructive help. They are already under the 1935–37 programme making great extensions. They are engaged on enormous commitments in frames and engines. They have not cavilled at financial provisions.[51]

Bristol's acquiescence (and eventual participation) in the shadow scheme ensured that the problems facing the company during the early years of rearmament were very different from those which faced Rolls-Royce. Five motor companies had to be trained in far more exacting engineering techniques than those to which they were accustomed, and immediate decisions had to be taken concerning the general production methods which the shadow factories would employ. The most important considerations here, again, were the total number of aero-engines to be manufactured, and the time limit which the government had imposed on scheme F.

Sir Herbert (later Lord) Austin, chairman of the Austin Motor Company and of the shadow engine group, was favourably impressed by Bristol's own approach to aero-engine production: 'The layout is simple, economic and easily controlled', he noted after a visit to Bristol's factory. 'Ideal in many respects . . . I do not think we can assume that we could reduce their present operation labour costs very much owing to the small number of engines to be made.' Austin also noted, however, that Bristol mainly used general-purpose machine tools. By employing special machinery he believed that higher levels of output could be reached, but the necessary investment would not be worthwhile when firm orders existed for just 1500 engines. Increased output in the long term might only be achieved at the cost of initial delays and of difficulties in maintaining complete interchangeability with engines manufactured by the parent firm.[52]

Austin's principal concern was that the Mark of engine might be changed after the first 1500 Mercuries had been delivered. Single-purpose machine tools could be used on particular components which were unlikely to be modified but flexibility would be necessary if major design changes were to be incorporated with the minimum of disruption, and this would be achieved only by the employment of standard types of machine. So it was decided that the shadow factories would follow Bristol production practice, introducing improvements only when they were justified by experience. Technical information was issued by Bristol on this basis but the envisaged production methods were sufficiently adaptable to allow line production and single-purpose plant to be employed at a later stage.[53]

The consequence of this decision was that the shadow factories duplicated the parent company's methods in almost every respect. Bristol advised the motor firms of the factory area required and of the size of the various departments. Operation times were based on those of Bristol's own machine department, suitably adjusted for the larger volume of shadow production. The machine tools employed in the shadow factories were those which had proved satisfactory at Bristol and the quantities of machine tools required were specified by Bristol and were based on Bristol machining times for each operation. In the absence of long-term information about the projected rate of shadow production, Bristol prepared machine tool plant balances to cover several different denominations of output, line, semi-line or merged production, as circumstances warranted. The monthly output of the shadow factories was not initially sufficient to demand the extensive use of line production or permanent set-up, but as output expanded full line production was introduced.[54]

As soon as a minimum of floor space became available at the new factories a nucleus of plant was installed so that test engines could be machined while the remaining construction work was completed. This enabled new staff and key production workers to gain experience of aero-engine manufacturing techniques and to 'prove' jigs, tools and other equipment. Close liaison between parent and shadow companies was vital to the success of the scheme. Representatives of the shadow industry paid frequent visits to Bristol's factory to observe the methods of production and to discuss problems with the parent company's experts; shadow industry personnel trained for special machining operations, assembly, inspection and testing, by working alongside Bristol's own employees and Bristol supplied the shadow firms with drawings and process sheets and with reference manuals dealing with every single aspect of Mercury engine manufacture from one end of the production line to the other.[55]

To supply such a substantial volume of technical information to the motor companies Bristol set up a department which became known as the shadow industry office. With the acceleration of rearmament its activities were continuously extended and it became an important medium for co-ordinating the shadow group as a whole, with responsibility for materials, bought-out parts and machine tools.

Shadow production operated under the direction of the senior executives of the participating motor companies, who formed the so-called Aero-Engine Committee. Matters of detail were dealt with by a subcommittee of the managers of the respective factories on which the Bristol shadow industry office and the Air Ministry were also represented. The managers' subcommittee, however, only met regularly while the scheme was being established, subsequent meetings being held to consider such matters as alterations to production programmes, factory extensions and the continuity of contracts. The main co-ordinating role was played by the manager of the Bristol shadow industry office, who was eventually to assume the formal role of co-ordinating engineer.[56]

SHADOW AERO-ENGINE PRODUCTION, 1937-39

The shadow industry initially received authority to produce 2000 Mercury 8 engines (1500 complete units plus spares) in November 1936.[57] At that time the Air Ministry did not foresee any further requirements for this particular model but it was unable to place orders for an alternative until the details of the next expansion programme had been finalised by the Cabinet. Authorisation was given to the industry to purchase materials for 450 additional engines in October 1937 but withdrawn a few days later, and by the beginning of the following year the DAP was having to take active steps to ensure that the shadow firms did not run out of work. 'I have already told factory managers unofficially that they will be well advised to slow up their production to that of the slowest member', Disney wrote at this time, 'and I have reason to suppose they are taking unofficial steps to do so.'[58]

Needless to say, this uncertainty resulted in growing concern within the industry itself. In March 1938, Rover's management committee noted that the

Shadow Factory Committee had written to the Ministry asking for a sanction to be issued for the second 2,000 sets to be manufactured . . . It was essential that this sanction should be granted now if the production of our parts were to proceed smoothly and economically.[59]

By this time, however, the *Anschluss* had forced a decision on the government. Two thousand Mercury 15 engines were ordered in the following week, and in May night shifts were organised and overtime was extended after the industry received instructions to increase weekly output from 50 to 75 engines.[60]

The technical assistance provided by Bristol enabled the motor firms to master aero-engine manufacturing methods relatively easily. Some problems were encountered but these were chiefly related to the structure of the shadow organisation. The participating companies were, of course, largely dependent on the supply of production drawings from Bristol.[61] The Mercury 8 was a new variant of an established design and, in March 1936, the only drawings to have been completed were of an experimental nature.[62] Bristol set up a separate drawing office for shadow engine work, and by November 1936 the company was anticipating that drawings of major components would be available by the end of December and those of minor components by the end of January.[63] However, the jig and tool drawing schedule subsequently slipped back to April

1937, and the drawing offices and tool departments of the motor firms themselves were brought in to clear the backlog.[64] The company most seriously affected was Humber, who in February 1937 required 3000 jig and tool drawings of which only 210 had been received from Bristol. This might not in itself have caused any long-term hold-up had Humber not also encountered technical difficulties in manufacturing the Mercury's blower unit. By December 1937, delays at Humber were threatening to disrupt the output of the entire shadow industry.[65]

The problem of balancing monthly output was ultimately solved when production at Humber gained momentum, and the target programme was reached by all the firms in October 1938.[66] There remained only the question of how to make up the cumulative deficiency. The shadow committee argued that the remaining arrears should be cleared by extending the Humber factory but the Air Ministry was initially opposed to any further capital expenditure.[67] Having received the support of the other shadow firms, however, Humber's chairman, William Rootes, immediately arranged for the factory to be enlarged and some months later the Air Ministry found itself facing a retrospective claim for £83,000.[68]

In the meantime, the programme of monthly output for the shadow group was increased, and plans were laid down to produce Bristol's 'Pegasus' engine during 1939. This proved the wisdom of the decision to opt for flexibility rather than true mass production but there were anxieties within the industry about the disruption which might be caused by manufacturing two types at the same time. These fears were not without foundation for although the Pegasus was very similar to the Mercury, it was equipped with a two-speed (rather than a single-speed) blower and the increased load thus fell on Humber.[69]

So the Air Ministry authorised further extensions and deliberately restricted the output of the other shadow firms.[70] Hence at Rover, where production was relatively advanced, it was recorded 'that the schedules as fixed by the Air Ministry would have the unfortunate result of reducing our output and this would necessitate the discharge of a large number of employees in the new year.'[71] The remaining deficiency at Humber was not rectified until July 1939, and only following the outbreak of war in September did the opportunity arise for the industry to fulfil its true potential, as Air Ministry demands increased from 80 to 100 and then to 150 engines per week.[72]

It is possible, however, to make too much of these early co-ordination problems. The arguments in favour of the multi-unit scheme were overwhelming in 1936 but the task of balancing early deliveries from several different factories, each making different components, would have taxed far more competent managers than those who presided over the shadow industry. There is no evidence that aircraft production was delayed by shortages of aero-engines at this time, and bearing in mind the novelty of the product it is surprising that more serious problems were not encountered and that production commenced as rapidly as it did. In fact the methods developed for transferring technical information from one industry to another proved highly successful and were duplicated when the second shadow aero-engine scheme was sanctioned in 1939.[73]

The overall efficiency of the shadow group was not reduced by Humber's difficulties, for the sharp increase in output between 1938 and 1940 was accompanied by a steady reduction in costs. In spite of the high wage rates paid in the Midlands, where the shadow factories were located, shadow costs were already comparable to those of Bristol by the time 500 engines had been produced (see Tables 18 and 19).

TABLE 18

Shadow Aero-engine Deliveries, 1938–39

	June 1938	Sept. 1938	June 1939	Sept. 1939	June 1940
Cumulative total	72	218	3,147	4,269	8,317
Monthly output, engines	42	96	342	340	747
Monthly output, spares		25	54	122	58

Source: AC 72/3.

TABLE 19

Mercury Production Costs: Bristol and Shadow Factories

	Batch	Average Cost Per Engine (£)
Shadow Industry	1st 500	1,915
	2nd 500	1,306
	3rd 500	1,197
Bristol	987	1,496
	887	1,486

Source: PRO AVIA 15/724.

ROLLS-ROYCE: PRODUCTION PLANNING, 1937–39

The uncertainty which surrounded the future of the rearmament programme at the end of 1937 caused as much concern to Rolls-Royce as it did to the shadow industry. The company's senior management was well aware that long-term RAF requirements for Rolls-Royce engines were increasing. A major programme of internal reforms had been implemented, substantial sums were invested in buildings and plant, and an extensive network of sub-contractors had been organised. The slow development of airframe production resulted, however, in the Merlin programme for 1938 being progressively reduced from 2220 to 1470, an output which did not absorb Rolls-Royce's maximum capacity. In January, Sidgreaves wrote to Swinton in the hope of obtaining larger Merlin orders for 1938 but the Secretary of State for Air could only reply that the future programme was under discussion at Cabinet level and that a decision was to be expected 'in the fairly near future'.[74]

Meanwhile, Rolls-Royce revised its policy towards shadow production. At the end of 1936 the Air Ministry had calculated that an annual output of 10,000 Rolls-Royce engines would be needed in wartime and had asked the company for its advice on how this target might be achieved.[75] It should be stressed that the Ministry was not considering the creation of additional Rolls-Royce capacity at this time.

Rolls-Royce did not initially see the shadow principle as a solution to the problem. The company was clearly reluctant to pass on its technical expertise to potential competitors but there were several other problems too. The Merlin was not, as yet, standardised. If the engine proved suitable for development then each standardised Mark would rapidly be superseded by more powerful variants. It seemed possible, however, that the engine might in any case have a relatively short service life, as more powerful types like Rolls's own Vulture and Napier's Sabre were under development. Either way, long production runs on standardised versions of the Merlin appeared unlikely and the engine did not therefore seem well suited to shadow production. To co-ordinate wartime production requirements with design and development, Rolls-Royce expressed a preference for concentrating all its resources, including motor car capacity, on aero-engine work, and for expanding the output of sub-contractors.[76]

The company's immediate proposal was that a department should be set up to examine the whole question of war production and that additional machine tools, jigs, patterns and equipment should be purchased and placed in storage in case of an emergency. It was anticipated that the cost of the scheme would be £100,000, and as it was concerned purely with war potential Rolls-Royce asked the Air Ministry to finance it. The DAP was agreeable to this but the AMSO's private secretary raised a number of objections, the principal one being that in wartime the normal capacity of several motor car firms would be available for Rolls-Royce engine production. So the proposals never reached the Air Council, and Disney had to content himself with an investigation of Rolls-Royce subcontractors which was undertaken by his own hard-pressed directorate during 1937. It was eventually concluded that sufficient extra capacity for war potential was available.[77]

Then, in September 1937, Hives changed his mind about shadow production. Rolls-Royce's objections to shadow schemes contrasted unfavourably with Bristol's readiness to co-operate with the state and the motor manufacturers. The Merlin II had now entered production and, with long-term requirements for the engine increasing, it was only a matter of time before a Rolls-Royce shadow was set up. The company was still unwilling to allow such a factory to be managed by the motor industry and it feared that the Air Ministry might at some future date ask the existing shadow group to build Rolls-Royce engines. Hives believed that Rolls-Royce itself should assume responsibility for management, and that it would be preferable for the necessary planning to begin sooner rather than later. Continued opposition to a Rolls-Royce shadow would, in these circumstances, have been detrimental to the company's own interests.[78]

In subsequent discussions with the Air Ministry, Swinton agreed to the idea of Rolls-Royce managing the prospective shadow factory. On 7 October, therefore, Hives presented a memorandum to the Rolls-Royce board in which he argued that 'if our engines . . . are not produced by shadow factories, it means that our business will necessarily be restricted'. On the following day Sidgreaves wrote to Disney stating that 'we are now quite willing for this company's products to be shadowed'.[79]

The paradox between the short-term over-supply and long-term dearth of Merlin engines was not lost on Swinton. At a progress meeting in July 1937 he voiced his concern about the lack of war potential for Rolls-Royce models, and it is a reflection of how serious the situation had become that, during the very same discussion, the decision was taken to re-engine the Handley Page P13/36 bomber with four Merlins in the knowledge that sufficient engines would be available for the number of aircraft then on order, which amounted to just 100.[80] Under emergency conditions, however, this ruling alone might have imposed a considerable strain on existing Merlin capacity. It is not, therefore, surprising to find similar misgivings being expressed by the Chief of the Air Staff in September.[81] Unfortunately, further consideration of the subject was delayed for six months during the protracted Cabinet discussions over the scope and expense of the next expansion scheme.

The approval of scheme L in April 1938 finally settled matters. The new plan increased the RAF's first-line fighter strength from 420 to 608 aircraft, adding significantly to Merlin requirements. The scope of both contracts and programmes was soon extended to March 1940, and Rolls-Royce was invited to submit proposals for expansion.[82] Despite agreeing to the shadow principle, the company initially proposed that all additional peacetime requirements should be fulfilled by subcontracting and by using existing car capacity for aero-engine production, thereby simplifying the task of management and supervision. Had this been possible, the shadow factory would only have been used in the event of an emergency. Rolls-Royce, however, failed to find an alternative location for its motor car division in Derby and it was therefore decided that a new factory should be built at Crewe to increase peacetime capacity by 150 engines per month.[83] By this time Rolls-Royce's total capital expenditure since the beginning of rearmament amounted to £750,000 and the company felt it could not reasonably be expected to spend any more. Financial liability for the Crewe factory was therefore accepted by the Air Ministry.[84]

Although the plant was initially intended to augment peacetime production and war potential, an extension was also planned as an insurance against the destruction of the Derby plant by enemy action. This naturally affected the basis on which it was planned. Every process and every source of supply used at Derby was duplicated as a strategic precaution. While planning and construction were in progress, however, rearmament was again accelerated (following the Munich crisis), and in January 1939 the Air Ministry informed Rolls-Royce that the entire factory was an immediate 'production necessity'.[85] Had this been the goal from the outset Rolls-Royce might never have embarked on the immense task of duplicating Derby's capacity, for the company would clearly have preferred to extend its existing subcontracting network instead. This would have allowed production to be increased much more rapidly in the short term.[86] The alternative of creating new or duplicate sources of supply, however, ensured that Rolls-Royce's capacity was much greater by 1941 than it would otherwise have been and there was much more potential for further expansion. As Britain was closer to a shortage of Rolls-Royce engines in 1941 than in 1939, the Air Ministry's insistence on longer-term planning was ultimately vindicated.[87]

It was hoped that Crewe would commence production in January or February 1939, but Hives soon concluded that 'the only possible way this can be achieved is by supplying most of the staff, supervisors and foremen from Derby, who know the job inside out.' A link-up between the new factory and Derby also seemed desirable to make the best use of new machinery. 'It will be from four to six months before we shall get any tools which could not be housed in the present vacant spaces in this [Derby] factory', Hives observed. 'The only way Crewe could be usefully used would be if it was considered a part of this factory.' He subsequently placed his ideas before the Rolls-Royce board:

The projected aero-engine factory at Crewe and the Derby works should be regarded as one unit for the purposes of administration and production since it will be of great advantage if parts, machinery, jigs and tools etc. can be transferred from one factory to the other to relieve hold-ups affecting output at either works.[88]

The construction of the new factory proceeded on this basis. Rather than merely copying the plant and organisation already existing at Derby, however, the layout at Crewe was planned 'in accordance with the latest conception of the industry', and new techniques were employed with the aim of reducing machining times. The plant also made greater use of specialised machine tools than Derby, and was more extensively tooled for production. Wages were paid on a straight piece-work basis, and the necessary rate-fixing department was established with further assistance from Production Engineering Ltd. Hives informed the Rolls-Royce board that he was 'introducing a number of economies which [he]would not be prepared to do in Derby at the present time for fear of dislocating production or giving rise to labour troubles'.[89] Crewe, however, was by no means a mass-production factory. As an insurance against the destruction of Derby, planning and layout at

Crewe reproduced something of the parent factory's flexibility so that new designs or variants could be introduced onto the production line.[90] An output of 35 engines per week was anticipated and, when the second section of the Crewe plant was approved in August 1938, this figure was doubled. Total Rolls-Royce capacity was raised from approximately 200 engines per month to 500 engines per month, single shift, and it was estimated that this output would be achieved in February 1940.[91] Average monthly delivery figures are shown in Table 20.

TABLE 20

Rolls-Royce Aero-Engine Production, 1939–41

(monthly average)

Year	1939	1940				1941			
Qtr	4	1	2	3	4	1	2	3	4
Derby	282	306	662	739	707	732	662	420	533
Crewe								428	518

Source: PRO AVIA 10/311 (production from Derby and Crewe was combined until the third quarter of 1941).

It initially proved difficult to finalise a clear programme of work for Crewe. The problem was the gulf between the short-term objectives of scheme L and longer-term estimates of wartime requirements. The government's aim was to create enough industrial capacity to satisfy the estimated wartime demand for Rolls-Royce engines but it was unwilling to commit itself to more engines than were required for the peacetime rearmament programme.

The programme which the Air Ministry eventually issued called for an additional 1900 Merlins and 50 additional Vultures by March 1940. The schedule envisaged a very rapid initial increase in monthly output but gave no indication of how Rolls-Royce capacity might be employed after existing contracts had been completed. As a result the company became increasingly concerned over the continuity of orders.

When instructions were received from the Air Ministry to proceed with the second section of the Crewe factory, Hives expressed his concern over Rolls-Royce's programme to Freeman. With an eight-month carry-over period required by the firm, the new monthly target of 500 engines would require orders in hand of at least 4000 engines at any one time. 'We must have a reasonable idea of the programme of work for this second extension', Hives declared. 'To justify the Crewe factory and the extensions at Derby full production must be maintained until the end of 1941.' He pointed out that under these circumstances cumulative orders would be far larger than those currently in hand and he evidently hoped for an indication as to whether Rolls-Royce would be safe in planning for such a high total.[92]

These arguments were received sympathetically by Freeman, but the absence of firm orders after March 1940 was still causing planning difficulties at Rolls-Royce some months later, and the Air Ministry was not authorised to extend the company's contracts beyond this date until February 1939.[93] It was, of course, the deterioration of international relations which finally allowed the long-term position to be clarified. In the meantime, planning for new Rolls-Royce engines such as the Vulture and the Peregrine remained much more uncertain, and even the Merlin programme was subjected to almost continuous revision. Airframe output was exceeding expectations; the demand for Merlin-engined fighters and bombers increased steadily. The consequence was that the Air Ministry changed Rolls-Royce's programme no fewer than six times in 1939.[94]

Increasingly the company sought greater production stability by planning output in accordance with information provided by the relevant airframe manufacturers.[95] Hives nevertheless resigned himself to the fact that quantity would always be subordinated to quality. 'We must be prepared', he wrote, 'to change our production programmes to meet the Air Ministry requirements at comparatively short notice.' This became the prevailing managerial philosophy at Rolls-Royce throughout the Second World War.[96]

THE EXPANSION OF AERO-ENGINE CAPACITY

In the aftermath of the Munich crisis, the Air Ministry carried out a complete review of the aero-engine industry's war potential. The proportion of multi-engined aircraft entering first-line service was increasing rapidly.[97] This facilitated the task of planning aero-engine production for it proved relatively easy to accommodate either the Merlin or Bristol's new Hercules in such aircraft. Plans were formulated on the basis of known information about the installation requirements for each type of engine but there was now scope for more flexibility in the way capacity was allocated than there had been at the beginning of rearmament. Hence Merlin capacity originally planned for the Wellington in 1939 was eventually assigned to aircraft like the Lancaster and the De Havilland Mosquito, while Hercules capacity was transferred to the Wellington; the performance of the Halifax bomber was greatly improved during the war by the installation of the Hercules in place of the Merlin, and both types of engine were installed into different versions of the Bristol Beaufighter.

The disadvantage lay in the escalating aggregate requirements for aero-engines, which ultimately convinced the Air Ministry that additional capacity was necessary. In Rolls-Royce's case the estimated deficiency of 400 engines per month was beyond the capability of Derby and Crewe, so a new and entirely self-contained factory was built at Glasgow where substantial manpower resources were available.[98] Rolls-Royce capacity was already being duplicated either at Crewe or by its subcontractors but it was decided that the Glasgow factory should be a second insurance against the threat of air attack and that practically the whole Merlin engine should be manufactured in Scotland independently of the existing Rolls-Royce organisation. To this end, provision had to be made for new foundry, pattern and tool departments, none of which were required at Crewe. Moreover, the preponderance of unskilled labour at Glasgow rendered the factory particularly reliant on up-to-date production methods and mechanisation. Machine tools were arranged and set for single-purpose operations, and the factory was planned on a flow production basis, with the minimum of rehandling. As a result, remarkable economies were achieved in the use of skilled labour, but only after an initial capital investment nearly double that required at Crewe for an equivalent monthly engine output (see Table 21).[99]

The expansion of Rolls-Royce capacity was mirrored within the Bristol shadow organisation. In April 1939, the Air Ministry asked Bristol and the shadow committee to erect two completely new factories to manufacture the Hercules engine.[100] There was, in fact, some concern within the Air Ministry that the design and development of the Hercules was incomplete at the time the new plans were initiated, and it is a reflection of the sense of acute urgency which characterised the Ministry's actions in 1939 that the risks were now accepted.[101]

The first of these factories was managed by a new shadow group comprising four motor companies (Standard, Daimler, Rover and Rootes) divided into two pairs so that each pair produced the complete engine. The second was managed by Bristol. Like Rolls-Royce's Glasgow works, these new factories were planned as mass-production units employing unskilled labour, flow production methods, and a high proportion of single-purpose machinery. The Bristol and shadow Hercules factories were initially planned to produce a combined total of 510 engines per month on a single-shift basis.[102]

TABLE 21
Rolls-Royce Crewe and Glasgow Factories:
Comparative Expenditure and Production

Factory	Initial Outlay	Planned Monthly Engine Output	Outlay Per Engine Per Month
Crewe	£2 million	303	£6,601
Glasgow	£5 million	400	£12,500

Source: Calculated from PRO CAB 102/53, pp. 27–30.

The aero-engine production plans laid down during the rearmament years (to which the Rolls-Royce shadow factories managed by Ford and Packard were added after the outbreak of war) established a network of manufacturing units organised in a manner typical of Britain's wartime aircraft economy as a whole. The Bristol and Rolls-Royce parent factories were planned to develop and produce the latest types or Marks of engine, some quantitative sacrifice being accepted.[103] Responsibility for bulk manufacture of the most successful variants was subsequently transferred to the shadow factories which, although lacking the research and development capacity of the parent companies, were much more suitable for mass production (see Figures 3 and 4). The Rolls-Royce organisation differed from that of Bristol only in the extent to which it subordinated quantity to quality. Rolls-Royce's Derby factory was, in Hives's words, 'a huge development factory rather than a manufacturing plant'. The Glasgow, Ford, and Packard factories were, on the other hand, planned for mass production from the outset but Crewe occupied an intermediate position.[104] Although it was better suited to production of large quantities than Derby, flexibility remained an important consideration there. This necessarily restricted the factory's productivity but enhanced its capacity to augment Derby's output of the very latest types of engine.

FIGURE 3

Bristol Aero-engine Production Organisation (1939 plan)

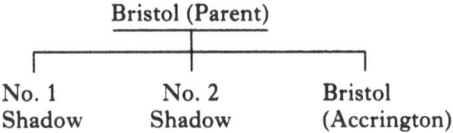

FIGURE 4

Rolls-Royce Aero-engine Production Organisation (1940 plan)

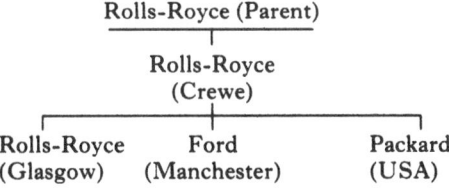

AERO-ENGINE PRODUCTION ON THE OUTBREAK OF WAR

The expansion of aero-engine production and employment during the rearmament years is illustrated by Table 22. Between 1935 and 1937, new capacity was established by enlarging existing aero-engine factories and by the creation of the shadow industry, but production developed slowly because of the change from the Kestrel to the Merlin at Rolls-Royce and balancing problems in the shadow industry. There was also considerable uncertainty about the extent of the government's requirements. Rolls-Royce produced fewer Merlins in 1937 than originally expected; yet early in 1938 the company was forced to cut production. It then received larger orders than it could immediately fulfil; but by the later months of 1938 planned capacity was again beginning to exceed foreseeable demands. The shadow industry was also directed to reduce its output at the end of 1937 and again at the beginning of 1939, and it did not realise its true potential until the outbreak of war.

Aero-engine production was occupying a false position in the aircraft programme. Under wartime conditions the aero-engine would assume primary importance, the industry would produce to capacity, and the availability of engines would exert a decisive influence over aircraft deliveries. In the rearmament years, however, there were no engine shortages, and the Air Ministry's attention was focused on the airframe sector. Only in 1939, when the supply of aero-engines began to cause official concern, was the long-term demand finally clarified.

TABLE 22

British Military Aero-engine Deliveries and Employment, 1936–39

Year	Deliveries Engines	Deliveries (m. hp)	Aero-engine Employment
1936	2,248	1.6	14,291
1937	3,440	2.7	16,936
1938	5,431	4.0	25,168
1939	12,499	8.3	30,983

Source: Production figures from *Statistical Digest of the War*, p. 155 (1936 figures April–December); employment figures from Weir 19/23A (figures for June 1936); PRO AIR 6/49; PRO AIR 6/58 (figures for August 1938 and April 1939). Employment figures for main contractors only.

Aero-engine production nevertheless expanded continuously throughout the rearmament years. In July 1939, the DGP reported that a threefold increase in the monthly output of engines had been achieved in the previous 12 months. He listed seven principal reasons for this: increased machine utilisation, increased working hours, increased production equipment, increased labour strength, increased area at parent factories, new area at shadow factories, and increased subcontracting.[105] There had also been significant gains in efficiency. An enormous programme of internal reorganisation had been implemented at Rolls-Royce, and the shadow industry had substantially reduced the unit cost of the Mercury engine. Further advances were to be expected during the war, given the higher degree of mechanisation adopted at the new Merlin and Hercules factories which the Air Ministry sanctioned in 1939.

AERO-ENGINE DEVELOPMENT ON THE OUTBREAK OF WAR

This chapter has been chiefly concerned with aero-engine production, but production matters cannot be completely divorced from the progress of design and development. During the rearmament years Rolls-Royce was engaged in several different development projects involving both the Merlin and other engines like the Peregrine, the Exe and the Vulture. The firm's evident relief at being able to concentrate production on the Merlin during 1938 was therefore tempered by the knowledge that several different types of engine might ultimately have to be produced side by side. 'We could not at the present time take on a production order for Exe engines', Hives informed Sidgreaves in March 1938. 'It looks as though it is certain we shall be producing Merlins and Vultures in parallel, and very possibly Peregrines . . . It means we have got to seriously reconsider our production facilities.' The following month he repeated that 'to have three types running concurrently will introduce quite a lot of difficulties'.[106]

The deterioration of the international situation added to these anxieties. 'If there was a war', Hives wrote in 1939, 'it would be obvious that the main production would be on the standardised and proved types of engines.' The advisability of introducing new designs into the existing factories was thus highly questionable. 'Our proposal is that it should be a definite policy of the Air Ministry that the plant for producing the standard engine, which in our case is the Merlin, should not be broken down to produce another type.'[107] One consequence of this line of thinking was that Rolls-Royce's development programme was cut back. Both the Exe and the Peregrine were abandoned during the early months of the war to concentrate as much technical effort as possible and to maximise output of the Merlin.[108] But there were dangers for Rolls-Royce in relying for quality on the development of existing models rather than the introduction of new ones, for the company's former competitor, Napier, was known to be designing the Sabre engine, the potential performance of which was far superior to that of the Merlin.

With hindsight it could be argued that the Air Ministry made a disastrous mistake by ordering the Sabre, for the engine proved one of the most difficult, costly and unrewarding of all pre-war and wartime development projects. When the decision to proceed with development was taken, however, much more seemed to be at stake than the technical promise of the engine itself. The fundamental tenet of Air Ministry selection policy was that quality could only be maintained by competition between different firms and the contraction of the aero-engine industry during the 1930s was therefore a continuous source of official anxiety.

In July 1935 the Secretary, Sir Christopher Bullock, argued that it was 'essential that Messrs. Napiers should be kept alive as a separate entity, in order to prevent the engine industry being constituted on too restricted a basis'.[109] In February 1937, Freeman argued that 'the loss of the experienced personnel making up the technical organisation [of Napier] would be a serious loss to the RAF', and in May 1939 he was again expressing his anxiety that not only Napier but Armstrong Siddeley were on the verge of leaving the aero-engine business. 'It was a most unhealthy position for the Air Ministry to be dependent substantially upon two firms only.'[110]

A development order had, however, been placed for six Napier Sabres during 1937, and in the same year plans were drawn up for a new Sabre-engined fighter, the Hawker Typhoon, to succeed the Hurricane and the Spitfire. The other aero-engine firms then began to encounter difficulties in developing the next generation of high performance reciprocating engines. The Rolls-Royce Vulture suffered repeated failures during 1939. The Bristol Centaurus was still in the most rudimentary stages of development and, as previously noted, production arrangements for the Hercules had to be made before the Air Ministry was certain that it would be successful. The decision to gamble on the Sabre was therefore based on two considerations: first, that the engine would provide a 'third string' in case the Vulture and the Centaurus failed, and, second, that it would maintain Napier in the aero-engine industry. On this basis the decision was taken shortly after the outbreak of war to erect a new factory at Liverpool with a capacity of 2000 Sabre engines per year.[111]

The renewed official interest in Napier caused considerable disquiet at Rolls-Royce. Following the decision to proceed with the Liverpool factory, Hives wrote:

We allowed Napier's to come into the aero-engine business after the war and build up a most lucrative business which should have been ours. If Napier's had really been awake we might have been suffering from that error today. Having now established ourselves as one of the two top aero-engine producers in the country, we have got to hang on to that position, even if it costs us money.[112]

Merlin development was therefore pressed forward, but the Vulture was still seen as the type most likely to counter the perceived threat from Napier.[113] Hives may, however, have harboured doubts about the Vulture[114] for, at the beginning of 1939, a very much simpler design project was initiated for a new 37-litre 1500 horsepower engine based on the Schneider trophy 'R' type which, unlike the Vulture, would fit the same installation as the Merlin. The attractions of the new project for war production were obvious, as Hives wrote to Freeman in February 1939:

The fact that this engine follows closely on the Rolls-Royce standard design, and the fact that we have an engine of such dimensions on which we shall shortly be running an endurance test, and also the fact that it is a similar engine to the 'R' engine, means that we are taking the minimum risk; far less risk than when jigs and tools are ordered for new aircraft.[115]

In short, the necessary technology for the engine, soon to be christened the Griffon, already existed. It had simply to be put into production.

Moreover, the new engine would give a much longer production life to existing airframes, particularly the Spitfire, and the challenge of an alternative Sabre-engined fighter might then be dispelled. This was obviously desirable from Rolls-Royce's viewpoint, but it was also in the national interest for Britain to avoid the costly transfer of aircraft factories from one type to another. At the end of 1939, Freeman told the Secretary of State that:

[i]n wartime when it is difficult to introduce new types of aircraft without a great falling off in production, it is essential that we improve the performance of the types which are already being produced.[116]

The Griffon project also attracted the attention of Supermarine, who were scheduled to change over from the Spitfire to the Beaufighter in 1941, when the Typhoon was expected to enter service. By October 1939 a scheme was in hand for installing the Griffon into the Spitfire, which promised to produce a fighter equal in performance to the Typhoon, but which was lighter and was equipped with a lower power engine.[117] The lighter airframe and the smaller engine would be more economical in terms of both man-hours and material.

It was characteristic of pre-war thinking that Rolls-Royce should have felt it necessary to offer a new engine to the Air Ministry and the aircraft manufacturers despite the company's professed intention of concentrating resources on fewer models. The inter-war years had witnessed remarkable advances in aero-engine performance, and in 1939 it seemed likely that this trend would be continued simply by designing larger engines. Identical assumptions were made by the other powers.[118] The reality, that reciprocating aero-engines were approaching the end of their economical development, would only become clear following the appearance of the jet engine in the next decade.

In the meantime, manufacturers made extravagant claims for engines like the Sabre, and although work continued on developing the Merlin the most important breakthroughs did not come until the following year. As late as April 1940 a memorandum on research and development prepared by Freeman for the Secretary of State for Air listed two Hercules variants together with the Sabre, the Griffon and the Centaurus, but made no mention whatever of the Merlin. In fact, the arguments used to support the Griffon project were much more applicable to the improved versions of the Merlin which emerged during the war because of the amount of capacity available for Merlin production. Within 12 months of the outbreak of hostilities, the realisation that Merlin development offered the most direct path to improved fighter performance caused both the state and the aero-engine industry to revise many earlier assumptions about the aircraft production programme.[119]

NOTES

1. Cairncross, *Planning in Wartime*, p. 15.
2. PRO AVIA 46/498, UK New Engine Deliveries in terms of Horsepower, April 1936 to May 1939.
3. After gaining experience with Rolls-Royce designs during the First World War, Bristol's chief designer, Roy Fedden, incurred a legal obligation not to work on liquid-cooled types.
4. PRO AIR 6/51, note by AMRD, 3 December 1937; PRO CAB 102/50, 'Reciprocating Aero-Engines, Design and Development', by J.L. Thorne (unpublished official narrative), sections 182 and 183. Only in the case of the Halifax bomber was the performance of an aircraft improved by substituting an air-cooled engine for a liquid-cooled engine, and this was due to the position of the power plant, not to the power of the engine.
5. PRO AIR 6/58, memorandum by AMDP, 12 September 1939. In the event the majority of Wellingtons were powered by the Bristol Hercules, and Merlins originally intended for the Wellington went to the Lancaster and Mosquito instead.
6. PRO AIR 6/33, S of S EPM, 29 March 1938.
7. PRO AIR 2/1322, statement of payments to aircraft industry on Air Ministry orders, 1923–1930.
8. Harvey-Bailey, *Hives*, p. 18.
9. Lloyd, *Years of Endeavour*, p. 232.
10. Information supplied to the author by Rolls-Royce.
11. Lloyd, *Years of Endeavour*, pp. 149–50, 250; Weir 19/23A, Air Ministry statement on employment in the aircraft industry, 30 June 1936; EEFAW, MSS 237/13/3, EEF statement of workpeople employed by aircraft manufacturers, 12 October 1934; Rolls-Royce board minutes, 24 January 1936. Part of this expenditure was for motor car manufacture.
12. Rolls-Royce board minutes, 4 October 1935.
13. PRO AIR 19/2, Sidgreaves to S of S, 25 October 1935.
14. Rolls-Royce board minutes, 28 August 1936.
15. *The Statist*, 22 May 1937.

16. Harvey-Bailey, *Hives*, pp. 27–8.
17. Lloyd, *Years of Endeavour*, p. 72.
18. RRA, Hives file, 'Policy', Hives to Lord Herbert Scott (chairman), Sir Arthur Sidgreaves (managing director), and Cowen (director), 18 January 1937.
19. Ibid.
20. Ibid; see also RRA, Hives file, 'Policy', Hives to Rolls-Royce board, 24 June 1937. Under both wage systems, standard times were fixed for every production operation and bonuses were paid when the standard times were undercut. Under premium bonus the saving was shared between the employer and the employee, whereas under piece work the entire saving was credited to the employee. The piecework system has recently been the subject of much criticism: see W. Lewchuk, 'The Motor Vehicle Industry', in B. Elbaum and W. Lazonick (eds), *The Decline of the British Economy* (Oxford, 1986), pp. 137–61.
21. RRA, Hives file, 'Policy', Hives to Rolls-Royce board, 18 January 1937.
22. Harvey-Bailey, *Hives*, pp. 27–8.
23. RRA, Hives file, 'Policy', Swift to Hives, 5 April 1937.
24. Ibid.
25. RRA, Hives file, 'Policy', report by Hives on aero-engine position, 27 April 1937.
26. RRA, Hives file, 'Policy', Hives to Rolls-Royce board, 4 October 1937.
27. PRO AVIA 10/12, Rolls-Royce brochure on Crewe factory, 12 January 1939.
28. RRA, Hives file, 'Policy', Hives to Rolls-Royce board, 4 October 1937.
29. Rolls-Royce board minutes, 21 May 1937.
30. PRO CAB 102/275, statement of employment at certain airframe and aero-engine firms, 1935–39 (prepared for the Official Historian).
31. RRA, Hives file, 'Policy', Hives to Rolls-Royce board, 24 June 1937; Rolls-Royce board minutes, 25 June 1937. The Hurricane and the Battle were, for example, extensively modified to accommodate the new engine.
32. RRA, Hives file, 'Policy', Hives to Sidgreaves, 13 January 1938.
33. Harvey-Bailey, *Hives*, p. 29.
34. Rolls-Royce board minutes.
35. RRA, Hives file, 'Policy', Hives to Scott and Sidgreaves, 6 April 1937.
36. RRA, Hives file, 'Policy', Hives to Scott, Sidgreaves and Cowen, 18 January 1937; Hives file, 'Copies of Important Documents', report by Hives on the aero-engine position, 27 April 1937.
37. RRA, Hives file, 'Policy', Hives to Scott and Sidgreaves, 6 April 1937.
38. RRA, Hives file, 'Policy', Hives to Scott, Sidgreaves and Cowen, 18 January 1937; Hives to Rolls-Royce board, 24 June 1937; Hives to Rolls-Royce board, 25 October 1937; Hives file, 'Production Engineering Ltd', memorandum by Hives, 20 September 1937.
39. PRO AVIA 10/12, notes by Hives for a meeting with AMDP, 6 September 1938.
40. PRO AVIA 10/12, Rolls-Royce brochure on Crewe factory, 6 January 1939.
41. Weir 19/21, memorandum by Hives, 15 November 1937; RRA, Hives file, 'Policy', Hives to Rolls-Royce board, 4 October 1937.
42. RRA, Hives file, 'Policy', note by Hives, 18 January 1937.
43. RRA, Hives file, 'Policy', Hives to Rolls-Royce board, 4 October 1937; memorandum by Hives, 13 January 1939.
44. RRA, Hives file, 'Policy', Hives to Scott, Sidgreaves and Cowen, 18 January 1937.
45. AC 79/2, draft manuscript on the history of the Bristol Aeroplane Company by C.H. Barnes, p. 139.
46. Between October 1933 and 1935 employment at Bristol increased from 2005 to 4223; by October 1939 the company employed 16,542 workers. See EEFAW, MSS 237/13/3, EEF statement of workpeople employed by aircraft manufacturers, 13 October 1933, 11 October 1935, 14 October 1939.
47. PRO AVIA 10/6, Fedden to AMDP, 6 September 1939. On the technical problems Bristol encountered in developing sleeve-valve engines see Gunston, *By Jupiter!*, p. 71.
48. Cross, *Swinton*, p. 162.
49. PRO AVIA 19/5, note of a meeting between the Air Ministry and the shadow industry, 29 June 1936.
50. PRO AIR 19/5, AMSO to S of S, 20 April 1936. A total of 4000 engines was to be pro-

duced by the end of March 1939; Herington (shadow aero-engine committee secretary) to S of S, 21 April 1936; notes of a meeting between the Air Ministry and the AESC, 29 June 1936.

51. Weir 19/23, S of S to Cooper, 16 March 1936.
52. PRO AIR 19/5, notes by Austin on a visit to the Bristol Aeroplane Company, 10 April 1936; AMSO to S of S, 20 April 1936.
53. AC 72/3, report by the manager of the Bristol Shadow Industry Office, 'History of Production of Bristol Aero-Engines by the Shadow Industry Operating under the Joint Aero-Engine Committee in Conjunction with the Bristol Aeroplane Company Limited', pp. 11, 53.
54. Ibid., pp. 6, 74–5.
55. Ibid., pp. 7–9, 114–15.
56. Ibid., pp. 2, 4, 27, 30, 39, 53; PRO AIR 2/2577, AESC meeting, 14 December 1936.
57. PRO AIR 2/2577, DAP to Austin, 26 November 1936.
58. PRO AIR 2/2577, DAP to Herington, 29 October 1937; DAP to Herington, 3 November 1937; DAP to AMSO, 16 February 1938.
59. Rover archive, MSS 226/RO/1/4/1, management committee meeting, 23 March 1938.
60. PRO AIR 2/2577, second DUS to AESC, 30 March 1938; Austin archive, MSS AU/1/1/2, board minutes, 4 May 1938; PRO AVIA 46/35, note by AMDP to the Air Council, 20 March 1939.
61. PRO AIR 2/2577, AESC meeting, 22 January 1937.
62. PRO AIR 19/5, Austin to S of S, 11 March 1936.
63. PRO AIR 2/2577, Stanley White (Bristol) to Herington, 2 February 1937; AESC meeting, 13 November 1936.
64. PRO AIR 2/2577, Clegg to DAP, 15 December 1936; AESC meeting, 14 December 1936; Cole (Standard Motors) to Herington, 26 January 1937; AESC meeting, 11 January 1937.
65. PRO AIR 2/2577, Rootes to Herington, 4 February 1937; AESC meeting, 1 December 1937.
66. AC 72/3, statement of engines, spares and accessory output from shadow industry, 1938–45.
67. PRO AIR 2/2577, AESC meeting, 28 April 1938.
68. PRO AVIA 10/156, ACCS meeting, 27 October 1938.
69. PRO AIR 2/2577, meeting between Austin, Bristol and Air Ministry, 10 June 1938. Two-speed engines gave two performance peaks at different altitudes, whereas single-speed engines achieved peak performance at higher altitudes. See Postan, Hay, and Scott, *Design and Development,* pp. 103–4.
70. PRO AVIA 10/156, ACCS meetings, 27 October 1938, 3 November 1938.
71. Rover archive, MSS 226/RO/1/4/1, management committee meeting, 18 January 1939.
72. Rover archive, MSS 226/RO/1/4/1, management committee meetings, 20 September and 31 October 1939; AC 72/3, statement of engines, spares and accessory output from shadow industry, 1938–1945.
73. AC 72/3, pp. 17–18.
74. RRA, Hives file, 'Policy', Hives to Rolls-Royce board, 4 October 1937; Hives to Sidgreaves, 13 January 1938; Sidgreaves to S of S, 14 January 1938; PRO AIR 19/2, S of S to Sidgreaves, 24 January 1938.
75. PRO AIR 2/1951, notes of a conference between the Air Ministry and Rolls-Royce, 7 December 1936.
76. Harvey-Bailey, *Hives,* pp. 29–30; RRA, Hives file, 'Policy', Hives to Sidgreaves, 13 September 1937; PRO AIR 2/1951, memorandum by Hives, 7 December 1936. There were also technical features which made Rolls-Royce designs unsuitable for shadow production. The Merlin had later to be completely redrawn before it could be produced by Ford and Packard during the war. See G. Mills, 'Ford and the Merlin', *The Archive,* Vol. 9, No. 27 (1991), p. 36; 'Building the Packard-Merlin', *The Archive,* Vol. 3, No. 22 (1989), p. 71.
77. PRO AIR 2/1951, DAP to AMSO, 8 December 1936; PS to AMSO to DAP, 2 March 1937; ADAP to DAP, 20 May 1937.
78. RRA, Hives file, 'Policy', Hives to Sidgreaves, 13 September 1937; Hives to Rolls-Royce board, 4 October 1937.

79. RRA, Hives file, 'Policy', Db to Hives, 14 September 1937; Hives to Rolls-Royce board, 4 October 1937; PRO AIR 19/2, Sidgreaves to DAP, 8 October 1937. The term 'shadow' was being used loosely here, because it was normally associated with capacity managed by companies from outside the aircraft industry. State-owned factories managed by the aircraft industry were generally referred to as 'agency' factories.
80. PRO AIR 6/30, S of S EPM, 13 July 1937. Two Vulture engines were originally specified for the P13/36.
81. PRO AIR 6/31, S of S EPM, 14 September 1937.
82. RRA, Hives file, 'War Emergency 1935–39', DAP to Hives, 18 March 1938; memorandum by Hives, 21 May 1938.
83. PRO CAB 102/53, 'Aero-Engine Production, Expansion of Capacity, 1935–45', by J.M. Embery (unpublished official narrative), p. 3.
84. PRO AVIA 10/151, ACCS meeting, 23 May 1938.
85. PRO AVIA 10/12, note of a meeting between AMDP, Hives and Sidgreaves; Rolls-Royce brochure on Crewe factory, 12 January 1939.
86. PRO AVIA 10/12, Hives to AMDP, 6 September 1938; RRA, Hives file, 'War Emergency, 1935–39', Hives to Sidgreaves, 30 December 1938; Sidgreaves file, 'Rolls-Royce Board No. 1', Hives to Rolls-Royce board, 25 September 1939.
87. On production in 1941 see RRA, Sidgreaves file, report by Hives, 26 February 1941; Hives file, 'Correspondence with Sir Charles Craven', Craven to Hives, 22 September 1941.
88. RRA, Hives file, 'War Emergency, 1935–39', Hives to Sidgreaves, 27 May 1938; memorandum by Hives, 2 June 1938; undated memorandum by Hives on Crewe factory.
89. RRA, Hives file, 'War Emergency, 1935–39', Ryce to Hives, 11 June 1938; Hives file 'Policy', memorandum by Hives, 13 January 1939; Lloyd, *Years of Endeavour*, p. 250.
90. PRO AVIA 10/12, Hives to AMDP, 16 December 1939; Lloyd, *Years of Endeavour*, p. 250.
91. PRO CAB 102/53, p. 27.
92. RRA, Hives file, 'War Emergency, 1935–39', notes by Hives for a meeting with AMDP, 8 September 1938.
93. RRA, Hives file, 'War Emergency, 1935–39', memorandum by Hives, 2 January 1939; Sidgreaves file, 'Rolls-Royce board No. 1', Hives to Rolls-Royce Board, 25 September 1939.
94. PRO AVIA 10/12, Hives to AMDP, 18 December 1939.
95. PRO AIR 20/2379, report by Sir Harold Howitt on a visit to Rolls-Royce, 25 January 1940.
96. PRO AVIA 10/12, Hives to AMDP, 18 December 1939.
97. PRO AIR 6/56, memorandum by AMDP, 20 March 1939.
98. PRO CAB 102/53, p. 40.
99. Lloyd, *Years of Endeavour*, pp. 197–9; RRA, Sidgreaves file, Rolls-Royce board No. 1, report by Hives, 26 February 1941.
100. AC 72/3, pp. 17–18.
101. PRO AVIA 10/6, Deputy Director of Research and Development (Engines) and DEP to AMDP, 17 January 1939; Fedden to AMDP, 6 September 1939.
102. Austin archive, MSS AU/1/1/2, board minutes, 26 April 1939; AC 72/3, pp. 17–18, 75; PRO CAB 102/53, pp. 15–16.
103. Memorandum by Hives, 15 March 1939, quoted in Lloyd, *Years of Endeavour*, pp. 189–90; PRO CAB 102/51, section 191, pp. 202-8, 224.
104. Lloyd, *Years of Endeavour*, pp. 189, 197.
105. PRO AIR 6/58, memorandum by DGP, 12 July 1939.
106. RRA, Hives file, 'Policy', Hives to Sidgreaves, 25 March 1938; memorandum by Hives, 4 April 1938.
107. RRA, Hives file 'Policy', memorandum by Hives, 19 June 1939.
108. PRO AVIA 10/12, AMDP to Hives, 30 August 1939; note of a meeting between AMDP and Hives, 14 October 1939.
109. PRO AIR 6/43, note by the Secretary, 1 July 1935.
110. PRO AIR 6/48, note by AMRD, 9 February 1937; PRO AVIA 10/6, note of a meeting between the Air Ministry and Bristol, 15 May 1939.

111. PRO CAB 102/50, section 106; Postan, Hay and Scott, *Design and Development*, p. 80.
112. RRA, Hives file, 'Copies of Important Documents', memorandum by Hives, 20 November 1939.
113. PRO AVIA 10/12, Hives to AMDP, 13 March 1940.
114. Harvey-Bailey, *Hives*, p. 51.
115. PRO AVIA 10/12, Hives to AMDP, 23 February 1939.
116. PRO AVIA 10/12, AMDP to S of S, 16 December 1939.
117. Vickers 410, Vickers-Armstrong (Southampton) works report, 23 October 1939.
118. In Germany, for example, an enormous technical effort was devoted to the Junkers Jumo 213 and the Daimler Benz 603. However, 'these engines . . . encountered design difficulties, with the result that great factory space was tied up unproductively for considerable periods of time.' See USSBS, Report 4, *Aircraft Division Industry Report (European Theater)*, p. 96.
119. PRO AVIA 10/357, AMDP to S of S, April 1940; PRO CAB 102/50, sections 131, 133, 137.

5

Manpower for the aircraft industry, 1935–41

In September 1935, the British aircraft industry employed 41,000 people, of whom 7000 were administrative and design staff and 34,000 were production workers. The fulfilment of the air rearmament programmes ultimately depended on increasing these figures. The task was always bound to be difficult. Few special measures were implemented by the government to improve the supply of supervisory and technical staff; the official policy was that rearmament should not interfere with ordinary trade, so compulsory transfers of manpower were out of the question. Instead the aircraft firms had to compete with civil industry for staff and labour. Many unemployed engineers had no experience of aircraft work, and the largest pockets of unemployment were in the north whereas most aircraft companies were located in the south. When they recruited unemployed or unskilled workers from other industries the aircraft manufacturers faced opposition from the engineering unions.

This chapter seeks to explain how the aircraft industry overcame these problems. Historians such as R.A.C. Parker have alleged that shortages of skilled manpower imposed a critical constraint on the progress of British rearmament, and Corelli Barnett has argued that such shortages seriously undermined the efficiency of Britain's wartime aircraft economy.[1] It is argued below, however, that effective manpower utilisation was no less important than manpower supply, and that in many respects the British aircraft industry made more efficient use of available manpower resources than its German counterpart.

THE SUPPLY OF MANAGERIAL AND TECHNICAL STAFF

The expansion of the aircraft industry created a demand for managers, draughtsmen and other technical personnel which substantially exceeded supply. 'Our greatest weakness is that we have not sufficient good designers, technicians and factory engineers,' Ernest Hives wrote, 'and in our expansion programme there is no scheme for improving this position.'[2] This sphere of policy was by no means ignored at government level, however. Expansion on the scale demanded by scheme F was recognised to be beyond the aircraft industry's existing managerial capacity and it was this consideration more than any other which resulted in the creation of the shadow industry. Further plans to employ managerial personnel from beyond the professional aircraft industry, in the event of war, continued throughout the expansion period, as firms like Vickers-Armstrong, English Electric and Metropolitan Vickers were introduced to aircraft manufacturing. Subcontracting, although primarily employed to tap labour resources, also helped to spread the burden of management.

Nevertheless, steps were also being taken to enlarge the professional aircraft industry which greatly increased absolute requirements for qualified supervisory personnel without making any provision to fulfil those requirements. Indeed, the Air Ministry generally saw effective management as a question of quality rather than quantity. The Ministry forced management changes on a number of firms but did not identify any overall failure of supply.

The same was true where technical staff were concerned. British aircraft design teams were smaller than their German or American counterparts, but considerably larger than those employed in other sections of the engineering industry.[3] Lord Weir thus found it difficult to accept that there was a shortage of technicians at the beginning of rearmament and he rejected out of hand proposals for pooling existing design capacity by rationalising the industry.[4] Rather than seeking to augment the supply of technical personnel, the Air Ministry looked to procedural changes to accelerate the pace of design and development, and most discussions with the industry on this subject were also confined to procedural matters. At the end of 1937 Freeman did suggest reducing the number of firms so that available design resources could be more efficiently employed, but any chance of implementing these proposals disappeared after the *Anschluss*.

Maintaining the supply of higher-level technical personnel for industry was, in any case, primarily the responsibility of the Board of Education rather than the Air Ministry. The Education Acts of 1918 and 1921 had aimed to extend Britain's technical training system, and some significant improvements were implemented during the 1920s but progress was consistently restricted, partly by government retrenchment which seriously interrupted technical education expansion programmes between 1931 and 1935, and partly by the opposition of industry to the daytime release of staff for training purposes.[5]

Towards the end of this period several suspended schemes were completed and a number of new projects initiated and in 1938–39 alone, 11 new junior technical schools were built.[6] By the later years of the decade, however, it was increasingly recognised by both government and industry that provision for technical education was inadequate and far inferior to that made in both Europe and the United States. British universities produced approximately 1000 graduate engineers per year, including about 100 aeronautics graduates,[7] but technical education was otherwise provided by practical 'on-the-job' training, combined with part-time, mainly evening, study. Of the 1,200,000 students enrolled in technical institutions in 1939, fewer than 50,000 attended full time.[8]

The Board of Education recognised that rearmament would impose significant additional demands for qualified technicians and managerial personnel, and it found that aircraft firms like Bristol, Short, Handley Page, Blackburn and Rolls-Royce were among the most enthusiastic advocates of improved technical education and training in the engineering industry.[9] The Board believed that proposals for improving technical education should not be too closely related to the short-term requirements of rearmament, for such an improvement was clearly essential to the country's prosperity under any circumstances.[10] As early as July 1935, however, complaints were received from Blackburn about the inadequacy of training facilities for production staff, and this led to discussions between the Board and the Air Ministry, while Blackburn approached the SBAC with proposals for a national policy for technical training in the aircraft industry. Owing to the stress of rearmament work, however, the industrialists failed to reach agreement over the thorny issue of daytime release, while the Air Ministry similarly argued that at such a time it 'could not be party to any attempt to bring pressure on the aircraft firms to secure daytime release'.[11]

This did not prevent the establishment of many important private and local initiatives which later received enthusiastic Air Ministry support, but it did rule out the implementation of a national

programme at the onset of rearmament. Such a programme would
have had no impact on the course of scheme C, and little on its
successor, scheme F, but it might have paid significant dividends in
the war years, when the shortage of technicians and managers
became acute. The sequence of short-term rearmament programmes and the absence of longer-term objectives proved to be the
greatest obstacles facing the advocates of improved technical education. Only following the outbreak of war would a series of training
programmes in technical colleges and universities significantly
increase the number of qualified engineers.[12]

How serious, then, was the shortage of managers and design
personnel? Where design staff are concerned, deficiencies were
encountered from the very beginning of rearmament, whereas the
shortage of supervisory personnel became increasingly pronounced
following the acceleration of the defence programme in 1938. In
December 1936, Vickers' technical staff, numbering 221, was one of
the largest in the airframe industry. Vickers' drawing office actually
employed 123, though Sir Robert McLean stated that he would
gladly have employed twice this number had the additional
draughtsmen been available.[13] In contrast, when Ernest Hives
visited the German aircraft firm, Heinkel, in 1935, he found no
fewer than 600 technicians on the staff.[14]

Such comparisons may be misleading, however, for a very substantial proportion of British aeronautical research was undertaken
by government bodies like the Royal Aircraft Establishment and the
National Physical Laboratory. The central role of these establish-
ments eliminated the need for the aircraft firms to maintain large
pure research staffs and the Air Ministry was spared the expense of
paying at least some of the research costs, overheads and profits of
20 different companies. The contribution of the government estab-
lishments to aeronautical design (see Table 23) must therefore be
borne in mind when considering the employment position in the
drawing offices of the Air Ministry's main contractors at the beginning of rearmament.

The statistics nevertheless disclose a general shortage of key
personnel – a shortage which was the principal cause of the delays
which beset the early stages of British air rearmament. In discussing
the slow progress of design and development with the Supply
Committee in 1938, both Frederick Handley Page and Richard
Fairey blamed their problems on the shortage of draughtsmen and
of drawing office section leaders and, writing in 1946, the Vice-

Chairman of Westland, Eric Mensforth, agreed. 'Staffs', Mensforth wrote, 'have undoubtedly been too small; an increase in their size would have allowed of a substantial reduction in the development period'.[15]

TABLE 23

Aircraft Industry Drawing Office Employment, 1935

Firm	Employment
Armstrong Siddeley	55
Armstrong Whitworth	161
Blackburn	96
Boulton Paul	65
Bristol (aircraft and engines)	328
De Havilland (aircraft and engines)	75
Fairey	218
Gloster	78
Handley Page	104
Hawker	109
Napier	141
A.V. Roe	114
Rolls-Royce (engines and motor cars)	264
Saunders Roe	138
Short	234
Supermarine	84
Vickers	176
Westland	57

Source: Weir 19/23A, figures for September.

No less serious were the shortages of production staff encountered after the implementation of scheme L. While planning the Rolls-Royce shadow factory at Crewe, Ernest Hives minuted that 'the question of staff is a great anxiety because there are not the men available', and six months later he had 'still to find quite a number of excellent works executive people for the first portion of Crewe'.[16] In the airframe sector, even the strongest firms encountered identical problems. Hence the Air Ministry's Director General of Production observed in 1938 that A.V. Roe's organisation was critically dependent on Roy Dobson, the general manager. Dobson was described as a man of 'great efficiency and drive' but he had apparently 'become seriously overloaded with a mass of detailed work'.[17]

In the absence of a state policy, the responsibility for improving this situation came to rest with the aircraft firms. Most companies operated apprenticeship schemes in collaboration with local colleges of further education and technical institutes and, by 1939, at least some training facilities specific to aircraft engineering existed in every major aircraft production location. Ten colleges offered Higher National Certificate (HNC) courses in aeronautics.[18] The close links between the firms and their local technical colleges are illustrated by the fact that, in 1940, all the students at Hull University's Department of Aeronautics were employees of Blackburn.[19]

For much of the inter-war period the industry favoured training schemes which combined on-the-job experience with evening classes, for which apprentices paid a premium. Increasingly, however, the trend was towards schemes for which no premium was payable. Ernest Hives had been particularly critical of Rolls-Royce's premium apprenticeship scheme when he became works manager at the end of 1936, and he soon abandoned it altogether in favour of a properly structured non-premium engineering course which included part-time release for HNC.[20] By 1940, four separate apprenticeship programmes were in operation which, in addition to the trade and engineering schemes, covered commercial engineering apprentices intended for future managerial positions and, for those with a university education, so-called 'probationer' engineers.[21] Similarly, Handley Page operated a non-premium scheme consisting of three years in the works and one on the staff. For the duration of the course apprentices attended classes at Northampton Polytechnic on a day-release basis.[22] Bristol and De Havilland augmented the supply of draughtsmen by establishing their own training schools, but extensive dilution, through the upgrading of apprentices and the introduction of women, was more common. In this way, design staff were eventually enlarged to a wartime average of approximately 300 employees per company.[23]

The progress of aircraft design and development was not, however, entirely a question of sheer numbers of staff. Much depended on the efficiency with which available resources were employed. In the early years of rearmament the Air Staff issued far too many specifications, but from 1938 new specifications were rationed to avoid overtaxing the industry's technical departments.[24] Handley Page managed to telescope the prototype and production drawings for their Halifax bomber, so economising on drawing office labour.[25] At A.V. Roe and De Havilland, self-contained experimental sections were established with their own staff, skilled labour, and machine tools, and these departments played a crucial role in reducing the development time of the Lancaster and the Mosquito. The Air Ministry's normal stipulation that a full set of drawings should be prepared for each design was relaxed for the Mosquito, and parts of the aircraft were developed entirely by De Havilland's experimental section in a form suitable for manufacture.[26]

After the onset of hostilities, the need to economise in the use of design resources ensured that less importance was attached to the introduction of new types of aircraft than to the modification of existing ones, which required only a fraction of the technical effort. Thus 330,000 man-hours were required to design the Spitfire Mark I, but only 620,000 man-hours were needed for all the subsequent 15 Spitfire Marks.[27] The same economic realities lay behind Rolls-Royce's decision finally to abandon the Vulture engine during 1941. 'We were certain', Hives wrote, 'that we could make a better contribution to the RAF both as regards quality and numbers, by developing the Merlin.'[28]

Shortcomings in the supply of managerial staff could also be overcome by using available resources more efficiently. Rolls-Royce was fortunate in that following the outbreak of war it was able to draw on its automobile division for engineers, draughtsmen, and jig and tool designers.[29] Hawker Siddeley also had sufficient resources to transfer staff within its organisation to those subsidiaries where the need was most urgent. Thus the Air Ministry's concern about the position at A.V. Roe persuaded Sir Frank Spriggs to allocate a number of young engineers to the firm, where they were trained to assist Dobson.[30]

Other companies did not have this option. In 1938 Richard Fairey therefore refused to manage a new shadow factory some distance from his Stockport works because of 'the strain which would be placed on the Stockport organisation if a further dilution were to take place', but he agreed to supervise another factory in Stockport on the grounds that 'the Stockport staff [could] handle double the present number of operatives employed, provided these were operating in buildings adjacent to the existing factories'.[31] Following the outbreak of war, Handley Page actually withdrew from two separate expansion schemes, largely because of the strain which they would have imposed on the company's resources.[32]

Efforts to prevent expansion from overwhelming the existing managerial resources of the aircraft industry were, however, frustrated by German bombing and factory dispersal in 1940.

Between 1938 and 1943 the number of factories employing more than 100 workers and managed by the main Air Ministry contractors increased from 45 to 323. The extent to which Britain's peacetime civil industry was converted to war production limited the number of managers available for transfer to professional munitions companies and the majority of staff was therefore promoted from within the firms themselves. To quote the SBAC:

the multiplication of managerial personnel required to cover new production units has been achieved almost exclusively by up-grading, i.e., making superintendents, assistant superintendents, and senior foremen into works managers; promoting charge hands to superintendents and foremen, and selecting their replacements from shop personnel.[33]

To summarise, the British aircraft industry's design and managerial capacity was maintained during the rearmament and wartime years by extensive upgrading of junior personnel, by a variety of short-term measures aimed at reducing the pressure on existing staffs, and by part-time training programmes provided by the aircraft firms, the local education authorities, and eventually the government. In comparison, German aircraft companies were supplied with more numerous and better-educated technical and supervisory staff.[34] In wartime, however, such advantages were of little value if they could not be translated into decisive quantitative and qualitative superiority, and only during the battle of France did Germany succeed in doing this. Subsequently, in spite of severe staffing difficulties, Britain's design and production record compared very favourably with that of Germany.

In the design sphere it is generally accepted that first-line interceptor fighters were of comparable quality in 1940, although the best German bombers were superior to most of their British counterparts, with the possible exception of the Vickers Wellington. By 1941, however, the British design programme was producing types like the Lancaster and the Mosquito, which would spearhead the strategic offensive against Germany throughout the remainder of the war, while the German design programme was on the brink of near total collapse, as aircraft like the Heinkel He. 177, the Junkers Ju. 288, and the Messerschmitt Me. 210 were plagued by technical problems and consistently failed to match performance targets.[35]

The paradox is partly explained by Britain's preference for developing more than one design for each military role, so that if one proved unsatisfactory another could be produced instead. However, Britain's relative success was also the result of a selection policy which gave priority to short-term projects that were expected to achieve quick results.[36] Ironically, this policy was dictated by the shortage of design capacity, for under wartime conditions the sheer volume of development and modification work either demanded by the RAF or proposed by the industry reduced the number of staff available for new, longer-term, ventures. German aircraft firms, on the other hand, devoted enormous resources to numerous (often unauthorised) design projects which did not reach fruition in time to make any contribution to the war. A similar approach could never have been contemplated in Britain because comparable design resources simply did not exist.[37]

Space does not permit a detailed comparison of the design features of individual aircraft. Where production is concerned, however, the British lead may be illustrated statistically (see Table 24). Compared with the later years of the war, British aircraft production was not particularly efficient in 1941, yet there was still a remarkable gulf between British and German production and per capita output. In other words, if the British aircraft industry was indeed handicapped by a lack of qualified managers this disadvantage was not reflected in the production statistics. The remainder of this chapter examines the approach to labour supply and utilisation problems which made these relatively high output figures possible.

TABLE 24

British and German Aircraft Employment and Deliveries,
1940–41

Year		Total Air Employment	Aircraft Deliveries	Airframe Weight (m. lb)	Aero-engine Deliveries
1940	Britain	973,000	15,049	59	24,047
	Germany	1,000,000	10,247	—	15,510
1941	Britain	1,259,400	20,094	87	36,551
	Germany	1,850,000	11,776	68	22,400

Source: R.J. Overy, *The Air War*, pp. 192 and 219 (1940 employment figures relate to April for Germany and December for Britain); German airframe structure weight figure from USSBS, Report 4, exhibit VI A.

LABOUR SUPPLY AND DILUTION

Labour for the aircraft industry

The supply of higher technical personnel received much less official consideration in these years than that of the rank-and-file work-force. The threat of skilled labour shortages seemed implicit in the government's decision not to transfer resources from civilian industries to the munitions sector, and in an attempt to alleviate the anticipated deficiency additional capacity was provided for engineering trainees at the Ministry of Labour's training centres.[38]

It was always clear that the centres could provide no more than a fraction of the munitions industry's requirements for skilled engineering labour. Some aircraft constructors were unhappy with the training provided by the centres, and the trade unions were strongly opposed to the introduction of so-called 'government trainees' when many of their members were unemployed.[39] Some labour shortages were thus inevitable. In general, requirements during the early stages of rearmament focused on skilled production workers but, in later years, increasing mechanisation created substantial demands for tool-makers and machine setters.[40]

However, it is questionable whether these bottlenecks were actually serious. Historians have made much of the skilled labour shortages encountered in the rearmament years,[41] yet the statistics indicate that the aircraft industry was not seriously affected by these shortages during the course of scheme F (see Table 25). Of the professional airframe and aero-engine firms' requirements for skilled labour for scheme F, 81 per cent had been fulfilled by 31 March 1937, and although the shadow factories were not included in these returns, they were being planned to use what was then a comparatively low proportion (30 per cent) of skilled workers.[42]

That the expansion the aircraft industry's labour force was undertaken very successfully throughout the rearmament period is also illustrated by the available figures for total airframe and aero-engine employment between 1935 and 1941 (see Table 26). According to the Official Historian, such shortages of skilled labour as did occur in the first years of rearmament were 'all local and, viewed quantitatively, unimportant'.[43] More serious deficiencies might have been experienced in 1938 had it not been for the recession which struck the general engineering industry in that year. The downturn in civilian demand allowed shadow firms like Austin and Standard to solve their labour problems by transferring workers from their motor car factories, while at Rolls-Royce aero-engine work was gradually introduced into the motor division.[44] Meanwhile, subcontracting provided another solution to any prospective labour shortages and was adopted as official policy by the Air Ministry in the autumn.[45]

TABLE 25

The Supply of Skilled Labour for Aircraft Production, 1937–39

	Skilled Labour Employed at 31 March 1937	Skilled Labour Requirements to 31 March 1939	Total
Airframe Firms	20,985	6,575	27,560
Aero-engine Firms	10,988	960	11,948
Total	31,973	7,535	39,508

Source: PRO AIR 6/49.

TABLE 26

Employment at Principal Airframe and Aero-engine Manufacturers, 1935–41

Year	Airframe Employment	Aero-engine Employment
1935	15,000	12,000
1936	36,000	14,000
1937	40,000	17,000
1938	65,000	25,000
1939	130,000	31,000
1940	173,000	80,000
1941	262,000	133,000

Sources: Aero-engine employment: Weir 19/23 A (September 1935 and June 1936), PRO AIR 6/49 (31 March 1937); PRO AIR 6/58 (August 1938 and April 1939); PRO AVIA 10/269 (June 1940 and 1941); 1935–39 figures for main contractors only; 1940–41 figures include main subcontractors.

Airframe employment: PRO AIR 8/196 (April 1935); PRO AVIA 12/194 (June 1936–39); PRO AVIA 10/269 (June 1940 and 1941); 1935–37 figures for main contractors only; 1938–41 figures include main subcontractors.

The outbreak of war brought increased competition for man-power resources, but following Dunkirk the aircraft firms were, for obvious reasons, given priority over the other munitions sectors for skilled labour. By August 1940 the government considered that an adequate nucleus of skilled workers had been absorbed by the industry and new labour allocation targets were established which assumed that virtually all the industry's additional demands could be met by unskilled workers. If skilled labour represented a problem, it was not a problem of supply but of distribution and during the following year the reallocation of existing resources between old and new factories became an important objective of the responsible government departments.[46]

Yet the key to the aircraft industry's success in fulfilling its labour requirements did not lie solely in the supply or distribution of skilled labour. There were, in fact, few aspects of aircraft production which could not be undertaken by semi-skilled workers and more mechanised production methods.[47] Manufacturing techniques had to be changed in order to reduce the industry's requirements for fully rated men, a process normally referred to as 'dilution'.[48] The main obstacle to dilution lay in the craft traditions of the engineering trade unions but, at the beginning of scheme F, Lord Weir advised against consultations between the government, the employers and the trade unions which, he felt, were 'likely to create a crisis type of atmosphere'. The Engineering Employers' Federation (EEF) also opposed the idea of national conferences with the Amalgamated Engineering Union (AEU) or the TUC, preferring the alternative of reaching local agreements with union representatives.[49] The aircraft industry itself was therefore to shoulder the main responsibility for dilution.

The aircraft industry and dilution

Given the prevailing concern over skilled labour shortages in 1935, it is perhaps surprising that neither the government nor the aircraft industry advocated any major expansion of the apprenticeship system to train large numbers of skilled workers over a period of several years.[50] Yet this was an entirely rational decision. Experience showed that skilled labour using handwork methods could take anything from ten to 50 times longer than semi-skilled labour using mechanised production processes to undertake comparable operations.[51] Rather than investing in long-term training, therefore, the aircraft firms had good reason to introduce semi-skilled or unskilled workers to jobs which had previously been undertaken by skilled men.[52]

There were several ways of doing this. First, companies could train their own employees on the job or establish their own training schools. Second, the industry could deskill the work itself by more extensive jigging and tooling or by subdividing operations. Third, semi-skilled workers could be upgraded to skilled work. All of these solutions were attempted during the rearmament years. Each entailed a measure of dilution by reducing the aircraft industry's dependence on apprenticed, unionised, labour, and thus it was inevitable that each should have encountered some opposition from the trade unions. Many union leaders believed that employers had reneged on First World War dilution agreements by failing to restore pre-war customs during the 1920s, and throughout the rearmament years the unions fought stubbornly to prevent the introduction of semi-skilled labour to work which had previously been monopolised by their members.[53]

It is sometimes suggested that employers used this union opposition as an excuse for their own laziness or conservatism.[54] Few aircraft industry managers were opposed to dilution in principle, however. As early as 1928, when the industry was contemplating the change from wood to metal construction, it was agreed by several leading industrialists 'that aircraft work is for the most part (perhaps 75–85 per cent) semi-skilled work' and that one result of the re-employment of woodworkers on metal 'was to prove that any metal job which the skilled woodworker could tackle was a semi-skilled job, inasmuch as the skilled woodworker had no craft training on metal'.[55] The moving force behind this resolution was Vickers Aviation. By September 1935, the proportion of skilled labour at Vickers' Weybridge factory amounted to just 27 per cent.[56]

The fact that the proportion at most firms was closer to 50 per cent in 1935 does not necessarily indicate a reluctance to deskill operations. In August 1935, the chairman of the EEF's Aircraft Manufacturers' National Technical Committee expressed concern that

the variation in practice . . . in regard to the classes of workpeople employed on the different operations . . . rendered the firms open to attack by the Trade Unions along the lines of comparison between the practice of one firm and another.

The committee concluded that 'aircraft manufacturers should endeavour to secure some degree of uniformity in regard to the various operations', but a subsequent survey revealed that there was in fact 'a fair degree of uniformity in the grading of operations' and that 'where a variation in practice obtains, this in most cases could be accounted for by . . . local conditions and local agreements existing in the general engineering trade'.[57]

The ratio of skilled workers to semi-skilled or unskilled workers also tended to vary according to the volume of output. Vickers was relatively prosperous in the early 1930s and had the resources to invest more heavily in plant and machinery than independent companies, some of whom produced only small numbers of aircraft in this period. Where contracts were too small to justify extensive jigging and tooling the corollary was that firms employed a high percentage of skilled labour. However, the larger contracts placed by the Air Ministry after 1935 entailed more jigging and tooling and more dilution. 'As the quantity of production grows, it is our intention to employ quite a good proportion of semi-skilled labour', the general manager of Gloster Aircraft informed the EEF in February 1938, and his counterpart in De Havilland's aircraft division similarly noted that 'where jigs are being used, semi-skilled labour only is required'. Boulton Paul likewise anticipated that 'the percentage of semi-skilled, boys and youths will increase as the necessary tools become available'.[58]

These opinions were shared by leading industrialists in the aero-engine sector. 'There is no mystery or fundamental difficulty in producing Rolls-Royce engines with unskilled labour', Ernest Hives wrote in 1939. 'It means a longer time in planning the production, more expense on jigs, tools and fixtures, the co-operation of the machine tool makers, freedom from alterations, and a longer time before production can commence.' This additional pre-production work had to be accepted as Air Ministry demands increased and the supply of skilled labour contracted. As Hives again noted, 'We could never hope to obtain sufficient skilled men at Crewe if we were using the same ratio as we are at Derby.'[59]

Finally, there are problems of definition. There is, of course, an important difference between a skilled man who has served a full apprenticeship and one who is simply being paid at the skilled rate. Many operations were graded as 'skilled' simply because of trade union restrictions, and deficiencies in the supply of skilled workers could thus be entirely artificial. When the acceleration of rearmament was under discussion following the *Anschluss*, however, the National Technical Committee pointed out that during the First World War

employers had the right to employ any class of labour on jobs which had previously been carried out by skilled men provided that the skilled rate of pay was offered in respect of the upgraded workmen . . . There appeared to be no objection to a similar arrangement being arrived at.

Wages were seen as 'a secondary consideration'.[60]

The extent to which this sort of upgrading was subsequently introduced is impossible to gauge with accuracy but the practice was clearly widespread. By February 1940, the payment of skilled rates to unskilled workers was described by A.V. Roe's Roy Dobson as 'the agreed bargaining system'.[61] Figures showing that 50 per cent of the aircraft work-force in 1939 was 'skilled' should therefore be treated with extreme caution.[62]

The opposition to dilution

The extent and nature of union opposition to dilution in the aircraft industry during the rearmament years is illustrated by the industrial disputes dealt with under the terms of national employer/union negotiating procedure. In the north, where the trade union movement was strongest, the resistance to dilution was more deeply entrenched than in the south, where the labour force was relatively fragmented. Southern firms, for example, had little difficulty employing government trainees. Bristol, Vickers and Gloster recruited considerable numbers during the 1930s, although these firms were careful not to employ trainees on work which was normally classified as 'skilled'. According to A.V. Roe, however, 'no [Manchester] employer could introduce these practices without inviting serious trouble from the Trade Unions', and an identical situation existed at Blackburn's works in Hull.[63] A strike was called at Fairey's new Stockport factory after the company recruited several men from government training centres. These 'trainees' were taken on because they were familiar with precision light engineering work, whereas such skilled engineering workers as were available in the Stockport area paradoxically required extensive training. The AEU nevertheless contended that trainees could not be engaged while skilled men remained unemployed.[64]

Aircraft firms had to be particularly careful to prevent the establishment of inflexible customs in the grading of operations. In the older factories this naturally proved difficult. At Handley Page the transition from small-scale to large-scale output led to an infringement of existing customs when machining work normally undertaken by skilled workers was allocated to a lower-rated man. The union maintained that the work had always been undertaken by skilled labour. Handley Page argued that although skilled labour had been used for small batches of parts where setting and operating was done by the same man, the work for larger batches had to be divided between skilled setters and semi-skilled operators. This was said to be the standard practice throughout the industry.[65] At one of the most important subcontractors, Dobson & Barlow of Bolton, the AEU attempted to establish Blenheim wing assembly as skilled work on the grounds that construction of the very first set of wings had been started by fully rated men.[66] A number of similar disputes relating to the assembly of metal-skinned aircraft convinced the employers that there was 'a definite drive on the part of the unions to establish this as skilled work', and resistance to such claims was successfully co-ordinated through the National Technical Committee during the early months of 1938.[67]

The dilution issue was complicated by the craft-based structure of British trade unions. In one wartime aircraft factory as many as 24 different unions were represented in the same department.[68] Certain unions, such as the sheet-metal workers' society, operated what the employers saw as particularly restrictive practices, and demarcation claims involving these organisations were strenuously resisted by the industry because of their implications for the progress of dilution. In March 1937, for example, sheet-metal workers at Fairey's Hamble factory claimed that their union had an exclusive right to undertake all metal-skinning work. The claim gave rise to an industry-wide survey in which the consensus was most accurately summarised by Roy Dobson. He wrote that

[t]he point is that metal skinning of aircraft would be an impossible job if we had to rely solely on tinsmiths to do it, and it is our intention to avoid the use of this class of labour entirely on such work. In some cases we are using fitters and in other cases entirely unskilled people.

The chairman of the National Technical Committee similarly expressed the view that 'we have to eliminate not only the sheet metal workers, but shipwrights and people of that nature, as this work is essentially semi-skilled'.[69]

The usual line taken by the industry when faced with the claims of one union against another was to deny that any single trade could monopolise a particular section of aircraft work, to suggest that the craft structure of the unions was inappropriate to aircraft production, and to allocate new names, which were not associated with traditional manufacturing trades, to the various operations.[70] In commenting on a dispute over wing production at one factory, an EEF director remarked that

the job was not a fitter's job, or a patternmaker's job, or a woodworker's job, but was one which had developed in the evolution of the industry – metal wing assembly, and the employer had the right to select the men who . . . were most suitable for the job, irrespective of the union to which they may belong.[71]

This was the outlook which governed the employment policy of Blackburn Aircraft when it associated itself with the shipbuilding firm, Denny, and built a new factory at Dumbarton, one of the few aircraft factories located in a 'distressed' area. Far more labour troubles occurred at this particular factory than at any other aircraft works, as Blackburn-Denny unwittingly inherited the numerous restrictive practices which had for many years characterised the Clyde shipyards. Of 37 disputes listed in the aircraft section of the EEF's case register in 1939, no fewer than 15 concerned Blackburn-Denny.[72] It is therefore worthwhile considering this case in some detail. The problems which arose were characteristic of those experienced by other manufacturers; their frequency was, however, unique.

In 1936 the aircraft industry was completely new to the Clyde and, as workers familiar with aircraft manufacturing were not available in the area, virtually the entire labour force for the Dumbarton factory was trained from scratch at a school established for that purpose.[73] Skilled workers were accepted from a variety of different trades, but Blackburn was anxious to avoid demarcation disputes and other restrictive practices and therefore adopted a revolutionary method of grading. All those engaged for the Dumbarton factory were simply categorised as 'aircraft assemblers' and were required to sign their cards as such, irrespective of their previous employment. Three grades, A, B and C, were applied to skilled, semi-skilled, and unskilled labour respectively, but the factory was planned so that jigs, tools and fixtures would be used much more extensively than at Blackburn's existing works. It was anticipated that the ratio of semi-skilled to skilled labour would be correspondingly higher. These arrangements were approved by the National Technical Committee in September 1937.[74]

Trouble started in December when the AEU registered objections to the employment of female machine operators and trainees and to the use of shipwrights on metal seaplane floats; the normal practice was to employ AEU fitters. Blackburn maintained that women were being employed as machine operators in many other engineering factories, and that company policy was to employ neither trainees, nor shipwrights, nor fitters, but aircraft assemblers. The various negotiation procedures failed to settle the dispute and, in March 1938, an EEF deputation visited the factory to investigate the situation. The deputation subsequently reported that Blackburn's attempts to overcome the traditional craft differentials had been carried too far and recommended that the firm should employ only fitters on skilled fitters' work. If more suitable employment could not be found for men from other trades, they should be discharged and offered semi-skilled work instead.[75]

Blackburn followed these instructions to the letter and inevitably incurred the wrath of the other unions, especially the boilermakers' society. A strike was called, and the boilermakers claimed exclusive rights to all riveting work (which was usually undertaken by fitters). The claim was rejected, but those boilermakers who had been discharged were reinstated as skilled workers in May. To placate the AEU the company promised that, where possible, these men should not be employed on jobs which were normally undertaken by fitters.[76]

This did not bring the factory's problems to an end. During the following months the shipwrights were subjected to threats and intimidation by the other unions which were attempting to enforce demarcation claims. In response, the shipwrights' union made its own demands for particular classes of work.[77] In October the sheet-metal workers laid claim to all fuselage shell plating and, in January 1939, the AEU demanded that fuselage assembly work on the Botha torpedo-bomber should be undertaken by their members alone.[78] Finally, the works manager was informed by the unions that 'all tradesmen employed as aircraft assemblers should be restored to the status of trade in which they served an apprenticeship [and that] all unskilled men employed on skilled jobs should be removed'.

Blackburn eventually agreed to abolish the grade B designation and to refer simply to 'semi-skilled operatives', but the management remained adamant about the need for flexibility on the part of the work-force. So the term 'aircraft assembler' was retained for skilled labour, and the demarcation question remained unresolved.[79]

The progress of dilution

Although Blackburn is a particularly extreme case, few aircraft firms were completely immune from dilution disputes between 1935 and 1939. The EEF case register for these years makes frequent references to the 'non-union question' and to union objections to the employment of trainees, semi-skilled labour and women on what was alleged to be skilled work. The dilution issue may also have precipitated demarcation disputes, as it did in the case of the Dumbarton factory. As the figures show, the industry still obtained most of the labour it needed for scheme F but when requirements for the next expansion scheme were considered by the National Technical Committee in March 1938 it was soon concluded that 'there was no possible solution unless dilution was permitted'.[80]

In May, Fairey's Stockport factory encountered more labour troubles, this time relating to the ratio of youths to skilled men, and a letter from Richard Fairey to Sir Charles Bruce-Gardner written some months later provides a further graphic illustration of the industry's difficulties. Fairey pointed out that while there were only 66 unemployed (and largely unemployable) engineers in Stockport, there were no fewer than 601 jobless cotton operatives. 'I can employ the lot and train them within a fortnight,' he wrote, 'but the unions will not allow it. In fact it is the unions' restrictions which stand between the industry and the labour it requires.'[81]

However, the progress of labour absorption could also be influenced by the ingenuity of employers, and it is clear that a considerable amount of 'hidden' dilution was taking place. This was not simply a question of subdividing operations, or paying the skilled rate to semi-skilled workers and passing on the bill to the government. In some instances dilution was physically concealed from the unions. When, in 1937, Armstrong Whitworth encountered objections from the AEU regarding the ratio of youths to adults in their Coventry factory, EEF investigations revealed very much higher ratios at both Bristol and Supermarine. Both companies insisted that their practices should not be disclosed to the unions.[82]

A.V. Roe also found that hidden dilution could solve many of their labour problems. Manchester was, as already noted, an area in which the unions proved particularly obstructive and Roe's general

manager, Roy Dobson, frequently adopted an uncompromising
stance when dealing with restrictive practices. In a dispute involving one of his principal subcontractors and the sheet-metal workers,
he dismissed the union's claims in typically forthright terms by
arguing that 'the work involved is not what I would call the working
of sheet-metal or a craftsman's job; that is anybody's job'.[83] Yet
Dobson's preferred strategy for A.V. Roe was to avoid these disputes altogether. This was achieved by decentralising production
into a number of branch factories in which new customs could be
established.

In 1935, A.V. Roe managed just one factory at Newton Heath in
Manchester, but by July 1938 four additional branches had been
set up at different locations throughout the city. The company's
first production work on an all-metal aircraft, the Blenheim, commenced in two such factories, allowing the firm to allocate semi-
skilled and unskilled labour to operations which might otherwise
have been claimed as skilled work by the unions at Newton Heath.
The EEF reported that in A.V. Roe's 'main works no metal wing
assembly [is] carried out and no semi-skilled men [are] engaged on
this work. In their Hamilton Street works, however, about 300
semi-skilled are employed on this work (Dobson does not want
union to be informed of this).' Sheet-metal work was dealt with in
the same way. All the skilled work for which fully rated men were
indispensable was transferred to the comparatively small Empire
branch works, while other operations were mechanised and allocated to unskilled workers in the main factories.[84]

It would clearly have been better from the production standpoint
had hidden dilution not been necessary at all. A.V. Roe's approach
reflected the strength of the engineering unions in Manchester;
instead of tackling the problem of restrictive labour practices head-
on, the company was obliged to concede to the unions at Newton
Heath the right to control the allocation of work. In the circumstances, however, hidden dilution was probably the only means by
which the supply of labour could be maintained and the process
therefore continued throughout the rearmament years in collusion
with Air Ministry. 'There is going to be considerable difficulty . . .
in getting the proper numbers of unskilled people and women on to
more or less skilled work', Dobson wrote to Lemon in February
1940:

I have been thinking around the Whitelands Road factory at Ashton . . . I
wanted to use this factory in the same way and for the same purpose which

we started Hamilton Street, that is to say, the introduction of completely unskilled labour on jobs which heretofore have been entirely skilled.

Dobson now suggested moving some of the work undertaken in Ashton to other factories, painting 'Air Ministry' on the door of the works 'so that it looked as if we were making a complete change over' and introducing machinery which could be operated by women or unskilled workers. 'I feel', Dobson concluded, 'that only by some such rather sweeping method we shall get away with it without considerable trouble with the unions.' The necessary expenditure was immediately approved by the Supply Committee.[85]

Such practices became widespread. As one union leader grudgingly admitted, 'whether we like it or not, dilution is spreading in the aircraft factories of this country today. Every man knows it. We have tried to control it, and we have failed to control it.'[86] The new capacity sanctioned by the Air Ministry from 1938 onwards marked a significant breakthrough in this respect, as it was planned in the expectation that skilled labour would not be available and that unskilled workers and more mechanised production processes would have to be employed instead.[87] Vickers, for example, intended that semi-skilled and unskilled workers should account for 85 per cent of the total labour force at their new Blackpool factory.[88]

Yet following the dispersal of the industry at the end of 1940, the Ministry of Labour noted that even

the old aircraft works . . . when manning their new dispersal units will not take their own transferee men from the old works. Apart from a nucleus staff they prefer to build up a new labour force on a basis of 50 per cent women [and] they welcome our co-operation in diluting their own works, thus enabling them to increase production and get the works running on a proper basis.[89]

Hawker and Bristol, like A.V. Roe, separated skilled sheet-metal work entirely from their main factories.[90] By the end of 1941 Short Brothers was employing 1268 people, including 743 women, as 'fitter assemblers' on sheet-metal components, while only 93 were employed as sheet-metal workers proper. At Castle Bromwich, sheet-metal fitting was removed from the main shops and put into new factories where semi-skilled 'pressings fitters' were used.[91]

The opportunity to establish different customs in new factories was also grasped by the aero-engine manufacturers. In December 1937 more than 60 per cent of Rolls-Royce's Derby labour force was rated as skilled, but by February 1941 only eight per cent of the

12,000 workers employed at the company's Glasgow factory were skilled men, and some of these were not from light engineering backgrounds.[92] Machinists' training schools were established at all three Rolls-Royce works, and at the end of 1940 it was estimated that 78 per cent of the machine operators at the Glasgow factory had been trained by the company.[93]

Progress would have been slower had the AEU not adopted a more flexible attitude towards dilution. There were several reasons for this change of mind. By 1939, unemployment among the union's members was falling and considerable dilution was taking place through subdivision and upgrading, which were often beyond the union's control. War seemed imminent, and there was a growing likelihood that the government would take special powers, which might have included industrial conscription. Any such development was viewed with trepidation by the AEU. A formal agreement with the employers, on the other hand, would increase the union's influence over dilution and improve its chances of restoring the established customs at a later date.[94]

In August 1939, an agreement was concluded between the AEU and the EEF, temporarily allowing the employment of unskilled workers on jobs previously done by skilled men. In the meantime a number of firms negotiated local agreements with union representatives permitting the replacement of men by women on certain operations, and in May 1940 the AEU had to acknowledge this trend by conceding a further relaxation of customs. During the following months the EEF reached dilution agreements with several other unions including the foundry workers, the electricians, and the sheet-metal workers.[95] The blanket priority over skilled labour which was granted to the aircraft industry during 1940 delayed the progress of dilution at some professional firms and the amount of development work undertaken by these companies inevitably created demands for skilled labour which were high in comparison with the newer mass-production factories.[96] Nevertheless, from 1941 onwards, the introduction of unskilled workers proceeded apace throughout the industry.

LABOUR UTILISATION

Recruiting labour was one problem; but increasing competition for manpower between the different sectors of the munitions industry and the armed services after 1939 raised questions as to whether existing labour resources were being used productively. Particular attention was focused on the high wages paid by some aircraft companies and on the general failure of the aircraft industry to increase per capita output during the first half of 1941. Indeed, it appeared to some of those in government departments responsible for labour allocation that the aircraft firms were wasting scarce resources which were urgently required elsewhere.[97]

However, it is difficult to accept such allegations completely given the intimate relationship between the government's own policies and some of the problems which faced the industry during mobilisation. These problems must be examined at the industrial level if their causes are to be properly understood.

Wages and earnings in the aircraft industry

One of the most complex industrial problems to arise from the transition from peacetime to wartime levels of output was the inflation of piecework earnings in the Midlands shadow factories. By the late 1930s many British engineering firms operated a piecework wage system comprising a basic hourly rate and a bonus, which was intended to act as an incentive to the work-force to increase output. Agreements between the EEF and the trade unions stipulated that workers of average ability should earn bonuses of at least 25 per cent above time rates. Piecework prices were normally set by rate-fixers after some production experience had been gained. Problems arose in the aircraft industry when the government placed large contracts for aeroplanes or components which had initially been ordered in small quantities. Under these conditions the original piecework prices invariably proved generous.

The piecework system has been the object of much criticism on the grounds that it weakened managerial control over production.[98] Less attention has been paid to the advantages of piecework. Companies favoured the piecework system because it acted, in effect, as a self-regulating mechanism of cost control. When a large volume of work was being undertaken, bonuses were high. Companies could afford to shoulder the higher wages bill and, as the rate of output increased, overheads tended to decline. On the other hand, when business was slack and a smaller volume of work was in progress, bonuses automatically fell. In this way firms could maintain a tight control over their labour costs.

In 1938, for example, Handley Page was in the process of changing over from the Harrow to the Hampden. In this transitional period sales actually fell slightly by comparison with the previous year but the labour force had to be enlarged prior to the commencement of Hampden production. Although this increase in employment amounted to 45 per cent over nine months, the piecework system enabled Handley Page to restrict the corresponding increase in their wages bill to just 12 per cent (see Table 27). In 1939, the expansion of output and piecework bonuses pushed up average hourly earnings from 19*d.* to 28*d.*[99]

TABLE 27

The Piecework System at Handley Page, 1937–39

Year/Aircraft	Sales (£)	Employment	Wages (£)
1937/Harrow	1,311,334	2,145	419,672
1938/Harrow and Hampden	1,261,265	3,115	468,460
1939/Hampden	3,408,898	4,935	1,089,442

Source: Handley Page accounts; EEFAW, MSS 237/13/3; PRO AIR 6/54.

The advantages of the piecework system were thus proven. By 1939, however, there were signs of increasing anxiety in the Air Ministry regarding the level of piecework bonuses being paid in certain sections of the aircraft industry and a seminal study of the subject was undertaken by the Director of Statistics and Planning, T.S. Smith, a Bedaux management consultant. Smith argued that piecework prices were being fixed too early, and that the acceleration of rearmament alone was leading to a fall in production times which allowed very high bonuses to be earned without commensurate effort from the work-force. Even in 1939 there was allegedly scope for some firms to increase per capita output by 25 per cent by paying lower bonuses.[100]

Smith recommended that rate-fixing should be delayed for as long as possible but it is likely that his analysis tended to oversimplify both the problem and its possible remedy. The period between the commencement of production and the establishment of new piecework prices was notorious for strikes, go-slows and other labour troubles,[101] and the task of rate-fixing under rearmament and wartime conditions was particularly complex. Until 1939 the aircraft companies had no knowledge of the government's long-term requirements and in the meantime piecework prices had to be fixed for aircraft and aero-engines which were not expected to remain in production for very long. There were, of course, many such designs. Only 310 Spitfires were ordered for scheme F in 1936, but by the end of war more than 19,000 had been produced.[102] Indeed, given the number of types which entered production in the rearmament years and then unexpectedly remained in production throughout the war, it is hardly surprising that difficulties arose in the operation of the piecework system. If anything it is remarkable that these problems were not more widespread.

The subject was considered by an Air Ministry committee in February 1940 but while it was recognised that, in some instances, high wage rates were being offered to attract skilled labour from other firms, there was as yet no evidence to suggest that wages were high in relation to those paid by companies engaged in other forms of large-scale light engineering such as automobile manufacturing. The problem was less one of wage rates than of bonus earnings in excess of the normal 25 per cent level. Even here, however, high bonuses were thought to result primarily from 'the combined effect of long hours, overtime, and Sunday work on standard rates and bonus earnings'.[103]

The EEF had drawn the government's attention to the problem of high earnings at State-owned shadow factories on a number of occasions but the supply departments had proved unwilling to intervene, fearing that any attempt to limit earnings might impede recruitment. There was no change of policy until late 1940, when the true extent of the problem became clear. In November, following discussions between the EEF and officials from Ministry of Aircraft Production (MAP), a special meeting of the National Technical Committee was arranged to examine the whole question.[104]

High earnings, it was agreed, were caused by piecework prices fixed during the early stages of production, when batches were small and workers undertook varying jobs. As output increased workers became more specialised, and had acquired greater skill and speed. Falling production times had then resulted in higher bonuses. At the shadow factories, piecework prices had sometimes been based on those already in operation at the professional aircraft firms, but the higher output of the shadow industry had inflated the earnings of its work-force. An instance was cited where a shadow factory was originally laid out to produce 50 units per week and where piecework prices were fixed for this level of production. In the course of time weekly output was increased to 200 units. 'There was no doubt that the men . . . were able to increase their speed on the job and this had been reflected in their earnings.' The rate-fixers' difficulties had been exacerbated by initial shortages of jigs, tools and materials. Time allowances had to be made for such factors when piecework prices were agreed but the tendency for these bottlenecks to clear as production gathered momentum created further scope for workers to increase their bonuses.[105] The committee concluded 'that any proposal to break down the piecework prices . . . at present existing in shadow factories would meet with considerable opposition . . . and that it would be extremely unwise to attempt such a proceeding'. Instead, it was agreed that

there should be a closer co-ordination between the initiating firms and the shadow factories . . . When new layouts were being contemplated, the shadow factories, before fixing piecework prices, should make every effort to contact the initiating firms with a view to obtaining advice on the matter.[106]

In the meantime, particular cases had to be treated on an individual basis. Agreements between the EEF and the unions provided for the amendment of piecework prices if rate-fixing mistakes were made, or if the means or volume of production were changed. Management had often been reluctant to use these procedures, but the agreements were now invoked by several firms in an effort to reduce earnings. For example, when the management of the Castle Bromwich Spitfire factory was transferred from the Nuffield organisation to Vickers-Armstrong it was discovered that not only had some piecework times been set incorrectly but that changes in production methods and in the scale of output were allowing bonuses of between 300 and 600 per cent to be earned in certain departments.

Vickers did not have the option of forcing down bonuses, as German companies did. Reductions had instead to be negotiated through the established procedures. The process took nine months but, in June 1941, an agreement was reached between the employers and the unions which allowed reductions to be gradually phased in, and in September Castle Bromwich reported that 'the difficulty experienced in the adjustment of prices . . . seems now to have been overcome. The average earnings have not greatly changed, but the output has appreciably increased.'[107]

By far the worst case in the industry, however, was that of the

Standard Motor Company's Beaufighter factory in Coventry. At the start of production Standard had never before constructed a metal airframe, and no information had been available from Bristol as to the likely manufacturing times. Under pressure from the MAP for deliveries, Standard issued piecework prices in January and February 1941, but during the following months the firm established a nightshift, tool and material supplies improved, more unskilled workers were employed and a system of line assembly was introduced. Despite several reductions in piecework prices, these developments led to a dramatic rise in the output of skilled production workers with consequences illustrated in Table 28.

TABLE 28

Piecework Earnings at the Standard Motor Company, 1940–42

	Earnings Per Hour (s.d.)	Bonus Per Cent
October 1940	2/6	
March 1941	3/5	200
September 1941	5/1	450
November 1941	5/3	550
February 1942	6/4	750

Source: EEFAL, P (5) 225.

It is worth emphasising that this situation only arose after the attainment of per capita output levels which greatly exceeded anything thought possible before the war.[108] The Standard figures reflect constant advances in productivity, and further progress in this direction was still being made despite the piecework problem. As Standard's manager declared, 'we shall not cease, and I am sure the men will not cease, efforts to find better means of producing . . . by additional equipment or change of process, so that there is a gain'.[109] Nevertheless, as the earnings of skilled workers at the factory were high even by Coventry standards, it was considered essential to reduce piecework prices. On this occasion the agreed negotiating procedures led to an overall cut of 40 per cent and a fall in production costs of 15 per cent.[110]

There were, however, alternatives to the established channels which were often more attractive to employers in the first instance. The introduction of new Marks of aircraft often gave managements the opportunity to negotiate lower rates. Thus the Castle Bromwich factory reduced piecework times for the Spitfire's wings by 800 hours when the Mark Vb was replaced by the Mark Vc.[111] Particular jobs could be 're-operated' by changing jigging and tooling arrangements, so allowing the employers to issue new piecework prices;[112] and the organisation of the aircraft industry into groups increased the scope for consultation between different firms producing the same type of aircraft.[113] Such measures were probably adopted far more widely than the formal arrangements for reducing high earnings.

Labour productivity, 1940–41

It was demonstrated in Chapter 3 that per capita output in the aircraft industry increased very considerably during 1939. In the early years of the war, however, such improvements were much more difficult to sustain. Indeed, between August 1940 and September 1941 there was no overall improvement in labour efficiency at all.[114] Throughout this period, production was subject to shortages of key supplies (such as machine tools, materials, and jigs and tools) which inevitably accompanied industrial mobilisation. Many firms were also in the process of changing over from one type of aircraft to another, during which time production levels tended to fall. The whole process was severely disrupted by German bombing and by the dispersal of aircraft factories. These problems will be discussed in more detail in a later chapter, but the statistics raise specific questions about labour utilisation in the aircraft industry which are more immediately relevant here. Why was it so difficult to improve per capita output between 1940 and 1941?

One important limitation on the performance of the industry was simply the reduction of working hours which occurred in the later months of 1940. The outbreak of war had been followed by an extension of working hours at most aircraft factories. Supermarine, for example, reported that, following the commencement of hostilities, day and night shifts of 63 hours per week had been instituted.[115] After Dunkirk, all departments worked a seven-day week so that the dayshift with overtime amounted to 70 hours while the nightshift was even longer.[116] At Hawker, too, labour worked twelve hours per day, seven days per week. Long hours of work provided the chief impetus behind the high production figures achieved by the British aircraft industry in 1940 but, in the words of a former member of Hawker's production staff, 'this could not be continued indefinitely [and] some relaxation followed'.[117]

Similar steps were taken by most firms. In November 1940 the National Technical Committee recorded the unanimous opinion that 'the present system of continuous working on seven days a week imposed a considerable strain upon the workpeople, and that it was essential that something should be done to relieve the situation'. It was initially decided that Sunday work should cease, but some months later the committee chose a six-day working week with factory closure alternating between Saturday in one week and Sunday in the next. This became the standard practice for the remainder of the war.[118]

To some extent, then, the stagnation of per capita output during 1941 reflected the necessary relaxation of emergency measures adopted during the Battle of Britain; otherwise the absence of any immediate improvement in productivity was itself a measure of the extent to which aircraft production had been de-skilled in the rearmament years. Very large numbers of unskilled and women workers were now recruited into the industry. Precise figures are not available but the proportion of female workers alone increased from 9.5 per cent in 1940 to 23 per cent in 1941, and in new factories, such as Castle Bromwich or Vickers' Chester works, the proportion was already above 30 per cent by September 1941.[119]

Recruitment is not, however, synonymous with absorption. Each new factory still required a nucleus of skilled labour which had to be transferred from other works.[120] The automobile companies overcame the problem by moving skilled workers from the first shadow aero-engine group to the second,[121] but this solution was not available to many other firms. The vast majority ran schools to train unskilled labour for repetition work,[122] but before unskilled workers could be effectively employed layouts had to be reorganised, factories had to be more extensively jigged and tooled, and production had to be divided into operations that could easily be learned.

All this took time. The government's own statisticians estimated in 1941 that absorption would take at least six months, but where new factories were concerned the process could be even more protracted.[123] Nearly two years after the decision to build the Rolls-Royce factory at Glasgow, Ernest Hives reported that the new works had not so far delivered an entire Merlin engine and that its chief contribution had been to produce parts for the company's other factories. In February 1941, productive hours equivalent to 41 Merlin engines per week had been worked at Glasgow but the factory had difficulty in producing just fifteen.[124]

There were also problems with labour discipline. 'Clydeside', Hives complained, 'is seething with communism . . . It is inevitable in any new factory that we collect a lot of rabble, who seize control of the shop stewards and shop committee.'[125] Similar difficulties were encountered at A.V. Roe's new Yeadon works, as Roy Dobson informed the MAP:

My main trouble so far as labour is concerned at Yeadon is the refusal of the workpeople to be properly disciplined, we having been constantly embroiled in the last couple of months in the difficult task of simply forcing discipline down their throats and trying to educate them to work-shop conditions. All this is due to the fact that practically all the labour at Yeadon is of the 'greenest' and most of it has never been in a workshop before and does not understand the necessities of production.

Dobson anticipated, however, that a marked improvement would take place during the following three months.[126]

That unskilled labour was successfully absorbed by the aircraft firms during 1941 can best be demonstrated by their productivity record in the following year. According to a memorandum prepared for the MAP's Chief Executive, between January and October 1942 output rose by nearly 50 per cent while in the same period employment increased by just 16.5 per cent.[127] The upward trend in per capita output continued until the later months of 1944.

CONCLUSION

At the beginning of rearmament it had been feared that aircraft production might be restricted by shortages of skilled labour. By 1941 the most exacting task facing the industry was not the acquisition of skilled labour but the absorption of unskilled labour. Over a period of six years the structure of the work-force had been transformed.

In the mid-1930s, the British aircraft industry, like its German counterpart, depended on a highly skilled labour force. British aircraft firms succeeded in fulfilling most of their requirements for skilled workers during the early stages of rearmament but they nevertheless endeavoured from the outset to increase the ratio of unskilled to skilled labour by mechanisation, upgrading and the subdivision of operations. Such methods helped the industry to overcome the severe shortages of skilled workers which were encountered after 1938. This, however, was by no means the only advantage gained from de-skilling aircraft production, for the new manufacturing methods resulted in productivity levels significantly higher than those associated with the equivalent skilled processes.

Although the aircraft firms encountered determined resistance from the trade unions, a considerable amount of dilution had already been achieved prior to the wartime relaxation of customs and this increased the scope for deskilling production during the war itself. Moreover, the new factories commissioned by the Air Ministry in the later rearmament years were planned to utilise an extremely high proportion of unskilled workers, and these factories added significantly to the output of the established industry when they entered production in 1941 and 1942. British employers did, of course, have difficulties with the piecework system and with labour absorption, but both of these problems were, in a very real sense, by-products of the industry's relative success in improving productivity and recruiting large numbers of unskilled workers. By 1941 the skills base of aircraft labour had been extensively undermined. The expansion of British aircraft production was, from then on, sustained by a largely semi-skilled and unskilled work-force.

NOTES

1. Parker, 'British Rearmament', pp. 318–20; Barnett, *Audit*, p. 132.
2. Hives to AMDP, July 1937, quoted in Lloyd, *Years of Endeavour*, p. 171.
3. During the war there was an acute shortage of draughtsmen in the airframe industry despite the fact that the industry employed 15 per cent of all mechanical engineering draughtsmen in Britain. See Mensforth, 'Airframe Production', p. 27.
4. PRO AIR 6/47, meeting between the Air Ministry and Vickers, 1 December 1936; Weir 19/13, Weir to S of S, 17 July 1935.
5. PRO ED 46/226, article published in *Education*, 28 February 1936.
6. Ibid.; PRO CAB 102/240, 'Education History: The Education Act of 1944', by S. Weifzman (unpublished official narrative), section 4, p. 44, footnote 4; p. 45, footnote 1.
7. PRO ED 46/257, Elliott to Wallis, 5 October 1944; conference on technical education convened by the Royal Aeronautical Society, 25 June 1943.
8. PRO ED 46/226, memorandum by H.B. Wallis, 13 January 1939.
9. Ibid.; PRO ED 46/226, Wallis to Tribe, 22 June 1939.
10. PRO ED 46/226, Stanhope (President of the Board of Education) to Inskip, 27 July 1938.
11. PRO ED 46/257, Secretary of the Board of Education to Secretary of the Air Ministry, 16 January 1936; minutes of a meeting between the Board of Education, the Ministry of Labour and the Air Ministry, 6 May 1936.
12. Barnett, *Audit*, p. 287.
13. PRO AIR 6/47, meeting between the Air Ministry and Vickers, 1 December 1936.
14. Harvey-Bailey, *Hives*, p. 28.
15. PRO AVIA 10/151, ACCS meeting, 29 April 1938; PRO AVIA 10/155, ACCS meeting, 8 September 1938; Mensforth, 'Airframe Production', p. 27.
16. RRA, Hives file, 'War Emergency, 1935–39', Hives to Sidgreaves, 27 May 1938; memorandum by Hives, 2 June 1938.
17. PRO AIR 6/54, memorandum by DGP, 5 September 1938.
18. PRO ED 46/257, statement by Board of Education on courses in aeronautics and

kindred subjects, March 1939.

19. HSA, Blackburn board minutes, 10 December 1940.
20. Harvey-Bailey, *Hives*, p. 23.
21. PRO AVIA 10/12, report by Director of War Production Planning, 24 February 1940.
22. EEFAL, A (1) 38, memorandum by the SBAC, 7 April 1938; PRO ED 46/257, memorandum by E. Savage, 17 June 1936.
23. Postan, Hay and Scott, *Design and Development*, p. 39; PRO AVIA 46/116, Postan's interview with De Havilland, 26 May 1943; Mensforth, 'Airframe Production', p. 27.
24. Postan, Hay and Scott, *Design and Development*, p. 38.
25. PRO AVIA 46/112, Postan's interview with Mr Volkert (designer at Handley Page), 12 May 1943. Design man-hours to the first flight of the prototype and to the completion of production drawings were identical – 325,000 hours. See Mensforth, 'Airframe Production', p. 27.
26. PRO BT 28/423, report by Sir Ernest Lemon, 12 August 1942; 'Personal history of S.D. Davies at A.V. Roe (Manchester), 1938–44'; PRO AVIA 46/116, Professor Postan's interview with De Havilland, 26 May 1943; note by De Havilland, 11 December 1941.
27. Postan, Hay and Scott, *Design and Development*, pp. 38–9, 161–2.
28. RRA, Hives file, 'Copies of Important Documents', memorandum by Hives, 8 October 1942.
29. RRA, Sidgreaves file, 'Rolls-Royce Board No. 1', Hives to Rolls-Royce board, 25 September 1939.
30. PRO AIR 6/36, S of S EPM, 1 November 1938.
31. PRO AVIA 10/154, ACCS meeting, 8 September 1938.
32. PRO AVIA 10/167, Handley Page to US of S, 24 January 1940; PRO AVIA 10/176, ASB meeting, 16 August 1940.
33. AC 70/10/58, memorandum by the SBAC, 8 January 1943. These figures did not include the shadow factories, nor the many dispersal units which employed fewer than 100 workers. Bristol, one of the largest and most dispersed of all, was also excluded.
34. Barnett, *Audit*, p. 150.
35. Overy, *Goering*, p. 193; *Air War*, p. 230.
36. Overy, *Air War*, p. 248.
37. Overy, *Goering*, pp. 183, 192; Postan, Hay and Scott, *Design and Development*, p. 38.
38. Parker, 'British Rearmament', pp. 307–8; Inman, *Labour*, p. 21.
39. EEFAL, A (1) 32, ACM NTC, 21 August 1935; Parker, 'British Rearmament', pp. 329–30.
40. Inman, *Labour*, pp. 28, 66.
41. Parker, 'British Rearmament', pp. 318–20; Barnett, *Audit*, p. 132; Inman, *Labour*, pp. 13–40.
42. PRO AIR 6/49, notes prepared for Secretary, 5 June 1937.
43. Postan, *War Production*, p. 96.
44. EEFAW, ACM NTC, 29 June 1938; PRO AIR 6/53 and PRO AIR 6/54, summary of reports from employment exchanges with regard to the Air Ministry's aircraft and aero-engine contractors, April and May 1938; RRA, Sidgreaves file, Rolls-Royce Board No. 1, Hives to Rolls-Royce board, 25 September 1939.
45. PRO AIR 6/58, memorandum by DGP, 12 July 1939.
46. Inman, *Labour*, pp. 31, 43, 48–57.
47. The industry believed that the proportion of skilled to unskilled labour could be reduced to 25 per cent. See EEFAW, ACM NTC, 17 March 1938. The term 'skilled' was normally associated with workers who had served a formal engineering apprenticeship although many of those classified as 'skilled' had not actually done so. The term 'semi-skilled' was normally applied to workers who were trained for specific tasks. As many aircraft assembly tasks like drilling, riveting or welding were extremely simple, a completely unskilled worker could become 'semi-skilled' in a matter of weeks or even days.
48. Weir 19/1, D of C to Weir, 12 June 1935.
49. PRO CAB 4/24, Defence Policy and Requirements Committee meeting, 27 January 1936; Parker, 'British Rearmament', p. 332.
50. The ratio of apprentice fitters to fitters actually declined between 1934 and 1935 from

1:10 to 1:15. See EEFAW, ACM NTC, 25 March 1936.

51. FBI archive, MSS 200/F/3/52/15/34, memorandum by the War Emergency Committee of the Institution of Production Engineers, 22 January 1941.

52. In April 1940 only one airframe factory, Rootes at Speke, was operating a long-term training scheme. See PRO AVIA 15/268, memorandum by Director of Labour Planning, 26 April 1940.

53. Parker, 'British Rearmament', p. 328.

54. Barnett, *Audit*, p. 157; Inman, *Labour*, p. 29.

55. EEFAL, A (1) 30, meeting of the aircraft sub-section of the subcommittee of engineering woodworkers, 28 November 1928.

56. Weir 19/23A, statement by the Air Ministry on employment in the aircraft industry, 30 September 1935.

57. EEFAL, A (1) 30, ACM NTC, 21 August 1935, 27 February 1936.

58. EEFAL, A (1) 46, McKenna (Gloster) to Campbell (EEF), 9 February 1938; Murray (De Havilland) to EEF London, 9 February 1938; Strickland (Boulton Paul) to the Secretary, EEF London, 2 February 1938.

59. Memorandum by Hives, 15 March 1939, quoted in Lloyd, *Years of Endeavour*, p. 190.

60. EEFAW, ACM NTC, 17 March 1938.

61. PRO AVIA 10/168, Dobson to DGP, 3 February 1940.

62. Inman, *Labour*, p. 33.

63. EEFAW, ACM NTC, 25 March 1936.

64. EEFAL, D (8) 42, local conference, 16 March 1936, 20 April 1936.

65. EEFAL, M (14) 13, memorandum by the EEF, 11 June 1936.

66. EEFAL, M (19) 69, report by Greaves (Dobson and Barlow), 13 January 1938.

67. EEFAL, A (1) 46, A.W. Grant (Bristol) to ACM NTC, 28 January 1938.

68. EEFAW, ACM NTC, 11 July 1941.

69. EEFAL, D (1) 145, enquiry by the EEF, March 1937; Grant to Campbell, 3 April 1937.

70. EEFAL, D (1) 139, local conference, 10 December 1930; EEFAW, ACM NTC, 5 January 1938.

71. EEFAL, D (1) 139, ACM NTC, 22 November 1935.

72. EEFAW, MSS 237/3/12/26.

73. EEFAL, A (1) 46, memorandum by the EEF, March 1938.

74. EEFAW, ACM NTC, 10 September 1937.

75. EEFAL, A (1) 46, works conference, 7 December 1937; memorandum by the EEF, March 1938.

76. Ibid.

77. EEFAL, A (1) 50, works conference, 8 July 1938.

78. EEFAL, A (1) 52, memorandum by EEF, October 1938; A (1) 55, works conference, 25 January 1939.

79. EEFAL, A (1) 55, Bumpus (managing director) to EEF North-West Association, 25 January 1939; works conferences, 27 January 1939, 31 January 1939; MacArthur (Blackburn) to EEF National Federation, 13 May 1940.

80. EEFAW, ACM NTC, 17 March 1938.

81. EEFAL, A (6) 41, Barlow (Fairey) to EEF Manchester Association, 14 May 1938; AC 70/10/67, Fairey to Bruce-Gardner, 27 October 1938.

82. EEFAL, A (1) 43, memorandum by EEF, 21 September 1937.

83. EEFAL, D (1) 45, local conference, 2 November 1938.

84. PRO AVIA 10/153, memorandum by A.V. Roe for ACCS meeting, 4 July 1938; EEFAL, M (19) 69, memorandum by EEF, 25 January 1938; interview with S.D. Davies, 30 April 1990.

85. PRO AVIA 10/168, Dobson to DGP, 3 February 1940; ACCS meeting, 10 February 1940.

86. TUC annual report, 1938, pp. 308–9, speech by Westwood of the ship-constructors and shipwrights union, September 1938, quoted in Parker, 'British Rearmament', p. 333.

87. Postan, *War Production*, p. 152.

88. PRO AVIA 15/268, memorandum by DLP, 26 April 1940.

89. PRO LAB 8/374, Egerton-Banks to Chegwidden (Ministry of Labour), 9 June 1941.

90. PRO LAB 8/217, report by the Ministry of Labour's south-eastern divisional office, 23 August 1938; AC 70/10/58, memorandum by the SBAC, 8 January 1943.

91. PRO LAB 76/28, 'Dilution in the Engineering Industry', by A.J. Corfield (unpublished official narrative), pp. 102–3.

92. PRO AIR 6/54, statement by the Air Ministry of employment at aircraft and aero-engine firms, 31 December 1937; RRA, Sidgreaves file, 'Rolls-Royce Board No. 1', Hives to Rolls-Royce board, 17 December 1940.

93. RRA, Hives file, 'Policy', memorandum by Hives, 20 September 1937; 'War Emergency, 1935–39', Hives to Sidgreaves, 27 May 1938; Sidgreaves file, Rolls-Royce Board No. 1, Hives to Rolls-Royce board, 17 December 1940.

94. Parker, 'British Rearmament', pp. 340–2.

95. PRO AVIA 10/12, reports by ADPWP, 24 and 26 February 1940; Inman, *Labour*, 31, pp. 57–9.

96. PRO LAB 8/374, Egerton-Banks to Chegwidden 9 June 1941.

97. ROBN 2/1/3, memorandum by the Programmes Department, Ministry of Aircraft Production, entitled 'The Measurement of Efficiency in the Use of Labour by Firms Manufacturing Aircraft', 23 June 1942.

98. Lewchuk, 'The Motor Vehicle Industry', in Elbaum and Lazonick (eds), *Decline*, pp. 139–40, 144–6.

99. Connolly papers, 'Survey of Relationship between Labour Effort, Aircraft Size, Jig and Tool Cost, and Rate of Output', by J.V. Connolly, 15 January 1940.

100. PRO AVIA 10/246, 'The Application of Wage Incentive in the Aircraft Industry to Facilitate Cost Control', by T.S. Smith (undated but late 1939).

101. PRO BT 28/423, report by Sir Ernest Lemon, July 1942; EEFAL, P (5) 188, local conference, 18 June 1936; S (9) 12, local conference, 2 June 1936; P (5) 222, local conference, 30 May 1941.

102. Mensforth, 'Airframe Production', p. 34.

103. PRO AVIA 10/168, report by the Departmental Committee on High Earnings in the Aircraft and Allied Industry, 7 February 1940.

104. EEFAW, ACM NTC, 26 November 1940 (on the creation of the Ministry of Aircraft Production see Chapter 7).

105. Ibid.

106. Ibid.

107. EEFAL, P (5) 225, EEF memorandum for central conference, 12 and 13 June 1941; PRO AVIA 15/2540, Dunbar (Vickers) to Craven (Controller General, Ministry of Aircraft Production), 20 June 1941; Vickers 200, 201 and 202, Castle Bromwich quarterly reports to March, June and September 1941.

108. This point was illustrated after the war by another industrialist using the example of a component which required 1014 man-hours to build at the start of production. Seventeen months later it required just 230 man-hours. See Mensforth, 'Airframe Production', p. 36.

109. EEFAL, P (5) 225, local conference, 19 June 1942.

110. EEFAL, P (5) 225, EEF memorandum for central conference, 9 and 10 July 1942; local conference, 19 June 1942; works conference, 15 June 1942; Standard archive, MSS 226/ST/1/1/7, board minutes, 20 November 1942.

111. Vickers 206, Castle Bromwich quarterly report to September 1942.

112. PRO AVIA 10/108, E. Walton (US Aircraft Production Board mission to Britain) to T.P. Wright (US Aircraft Production Board), 24 July 1943.

113. On rate-fixing in the Lancaster group see Vickers 212, Castle Bromwich quarterly report to March 1944.

114. ROBN, 2/1/3, memorandum by the Programmes Department, Ministry of Aircraft Production, 23 June 1942.

115. Vickers 194, Supermarine quarterly report to September 1939.

116. Vickers 197, Supermarine quarterly report to June 1940.

117. Letter by A.E. Tagg (formerly of Hawker Aircraft Ltd.) to the author, 12 December 1990.

118. EEFAW, ACM NTC, 26 November 1940, 13 May 1941.

119. Inman, *Labour*, 80; Vickers 202, Castle Bromwich and Chester quarterly reports to September 1941.

120. Inman, *Labour*, pp. 43–4.

121. AC 72/3, 19.

122. For example, such schools were operated by Rolls-Royce, A.V. Roe, Fairey, Blackburn-Denny and English Electric. See PRO AVIA 15/268, DPWP to DDGP 1, 21 March 1940; PRO AVIA 10/165, note by DGP, 9 December 1939.
123. ROBN 2/1/3, memorandum by the Programmes Department, Ministry of Aircraft Production, 23 June 1942.
124. RRA, Sidgreaves file, 'Rolls-Royce Board No. 1', Hives to Rolls-Royce board, 17 December 1940; Harvey-Bailey, *Hives*, p. 50. For a more detailed discussion of the labour problems encountered at Rolls-Royce's Crewe and Glasgow factories see I. Lloyd, *Rolls-Royce: The Merlin at War* (London, 1978), pp. 24–33.
125. RRA, Sidgreaves file, 'Rolls-Royce Board No. 1', Hives to Rolls-Royce board, 17 December 1940.
126. PRO AVIA 15/159, Dobson to Sir Archibald Rowlands (Permanent Secretary at the Ministry of Aircraft Production), 10 April 1942.
127. PRO AVIA 9/70, memorandum by the Controller General to the Chief Executive, 9 November 1942.

6

Finance, contracts and prices in the aircraft industry, 1935–41

Between April 1936 and March 1942 the government spent £204 million on buildings and plant for aircraft production.[1] But the aircraft themselves were far from inexpensive. In the same period, production expenditure on aircraft and associated products came to more than £1.6 billion.[2] With the government injecting such vast sums of public money into a private industry, the issue of the financial relations between the state and the aircraft contractors assumed a new importance.

At one level this was simply a question of financing the enlargement of fixed and working capital. Whether this was undertaken by private enterprise or by the state depended on the extent to which the financial requirements of rearmament exceeded those of the pre-expansion period, or those which the industry expected to be necessary following the completion of the defence programme. The mid-1930s witnessed a remarkable boom in aircraft investment as new capital was sought to finance schemes C and F but, as the demand for aircraft grew even further during 1938 and 1939, the state was called on to assume an increasingly important role in financing production.

At the contractual level, however, the relationship between finance and production was more complex. As the state played no part in factory management, the aircraft firms exercised complete control over investment decisions and over the way in which capital was employed. Even when the government began to invest directly in factories and plant in 1938, detailed decisions regarding the utilisation of capital were still taken by the industrialists. Contractual policy did, however, provide a direct way in which the state could encourage expansion and influence production methods. Moreover, the state had a political obligation to ensure that aircraft were manufactured economically and that the prices charged for them were not excessive. Throughout the rearmament years, repeated attempts were made to reach a financial settlement which performed these various functions.

THE AIRCRAFT BOOM

The collapse of the Geneva disarmament conference together with factors such as the government's defence programme, the growth of civil air transport, and rapid technological development stimulated an unprecedented demand for aircraft shares in the mid-1930s. At the beginning of rearmament in 1934 there were relatively few public aircraft companies. Such substantial concerns as Armstrong Siddeley Development and Bristol were privately owned, while Vickers and Supermarine were subsidiaries of Vickers Ltd and did not publish their own accounts.[3] So when the aviation industry began to attract the attention of investors, they found only six established public aircraft companies with a total share capital of just £3.8 million.[4] Share prices soared. Little heed was taken of the fact that Fairey's profits had fallen by 42 per cent in 1933 or that Handley Page and Napier had actually made losses. The banking firm, Lazard Brothers, which acquired a substantial stake in Napier, appears to have been completely ignorant of the company's lack of technical success until 1937.[5]

TABLE 29

Aircraft Industry Share Prices, 1933–34

Firm	1933 Low (s)	1934 High (s)	Share Capital (£) 1933–34
Hawker (Ord. 5s)	10	29	787,000
Fairey (Ord. 10s)	16	35	500,000
Rolls-Royce (Ord. £1)	44	111	839,000
De Havilland (Ord. £1)	14	66	400,000
Handley Page (Pref. 10% 8s)	6	17	206,000
Napier (Ord. 5s)	3	15	1,028,000
Total Share Capital			3,760,000

Source: *The Statist*, 30 March 1935.

Table 29 illustrates the contrast between the lowest share prices in 1933 and the highest in 1934. In the following years both Fairey and De Havilland doubled their capital by floating new issues while Rolls-Royce and Handley Page capitalised reserves and, in Handley Page's case, additional sums from their patents realisation account. However, Napier's relatively large capital was not increased at all. The total share capital of these firms (excluding Hawker, which became part of the Hawker Siddeley group) amounted to £4.6 million on the outbreak of war.[6]

TABLE 30

New Public Aircraft Companies Formed Between 1933 and 1936
(Air Ministry Contractors)

Firm	Authorised Capital on Formation (£)	Issued Share Capital, 1939 (£)
Hawker Siddeley	2,000,000	6,000,000
Bristol	1,200,000	3,900,000
Blackburn	630,000	1,200,000
Boulton Paul	250,000	500,000
Westland	250,000	400,000
Short	150,000	500,000
Total	4,480,000	12,500,000

Source: *The Statist.*

The older public companies were joined by the other established Air Ministry contractors between October 1934 and April 1936 (see Table 30). One of the most important firms in this group was the Bristol Aeroplane Company. In June 1935 Bristol's net assets were valued at £1.2 million, and profits had averaged £150,000 in the five years prior to flotation. The company's stock was thus extremely attractive to investors and, when it was announced that 360,000 of Bristol's ordinary 10s. shares had been acquired by the introducing brokers at 37s. 6d., so much interest was generated in the issue that, on the day after dealing commenced, prices stood at 56s. 3d.[7] Share premiums from the new issues of 1935 and 1936 amounted to £900,000.[8]

Hawker Siddeley was formed in the following month to acquire the whole of the ordinary capital of Armstrong Siddeley Development (comprising A.V. Roe, Armstrong Siddeley, and Armstrong Whitworth) and half the ordinary capital of Hawker Aircraft, a subsidiary of which was the Gloster Aircraft Company. Hawker's remaining capital was absorbed by Hawker Siddeley in the following year. The new company's 5 s. ordinary shares were initially marketed at 15 s., but by February 1936 their price had risen to 34 s. 6 d. In the aftermath of flotation the company's share premium account was credited with more than £800,000.[9] The story of Bristol and Hawker Siddeley in the rearmament years is one of continuous expansion. By 1939 the share capital of these two concerns amounted to nearly £10 million, while a further £5.8 million had been raised through debentures and loans.[10]

By comparison, the other companies in this group operated on a relatively modest scale. Westland and Boulton Paul were originally the aircraft sections of manufacturing companies whose chief interests lay outside aviation, while Short Brothers had derived a substantial portion of its income from manufacturing bus bodies. Although firms like Westland had made large profits in the 1920s, aircraft work had not been remunerative in the years preceding rearmament and these companies initially viewed the prospect of flotation on the Stock Market with some trepidation. Blackburn recognised in June 1935 that additional capital would be required to finance expansion but went to considerable lengths to obtain it through private channels before finally deciding to go public.[11]

The process of flotation was not without its difficulties. When Westland's shares were offered to the public by their parent company, Petters Ltd, in 1935, only £75,000 of Westland's total capital of £250,000 was subscribed in cash. The balance of the capital, consisting of shares with a nominal value of £175,000, was issued to Petters as a purchase consideration. Petters then sold £125,000 worth of these shares to the public at a premium of 40 per cent but the proceeds of this sale went to Petters rather than Westland.[12] Only the strongest possible pressure from the Air Ministry persuaded Petters to adopt a more responsible attitude towards Westland's finances.[13]

Many new aircraft companies emerged alongside the established Air Ministry contractors. These firms usually had licences to build foreign designs but the majority were probably hoping to undertake subcontracts for the rearmament programme. Their shares proved no less popular than those of the established firms. Indeed, Airspeed, General Aircraft and Phillips & Powis became public companies before several of the older concerns.[14] One newcomer was Parnall Aircraft, which eventually manufactured the Fraser-Nash gun turret. The first shares issued by Parnall were oversubscribed 14 times.[15] Another was the Aeronautical Corporation of Great Britain (ACGB) which was successfully floated in May 1936, ostensibly to produce the 'Aeronca' aircraft, which was said to have been approved by the Air Ministry as a trainer. Profits of 20 per cent on issued capital were confidently predicted.[16] In 1936, the Air Ministry found that 17 firms from beyond the established circle of contractors were available for rearmament work.[17] Nine were public companies (see Table 31).

The City was deeply disappointed by the earning capacity of these concerns. One analyst referred in 1937 to 'the folly of excessive optimism in the investment sphere and the utter recklessness with which money was subscribed for shares in new undertakings not on grounds of past achievement but merely on the company promoter's roseate picture of things to come'.[18] Neither ACGB, British Marine, nor Aero Engines survived for very long,[19] and three firms, Pobjoy, Phillips & Powis and Airspeed, soon lost their independence. Phillips & Powis was acquired by Rolls-Royce, and Swan Hunter bought a controlling stake in Airspeed in 1934. This was, in turn, sold to De Havilland in 1940.[20] As subsidiaries, however, these firms became two of the most important manufacturers of training aircraft in the industry.

TABLE 31

New Public Aircraft Companies Formed Between 1933 and 1936
(Non-Air Ministry Contractors)

Firm	Design (D), Patent (P), Licence (L)	Share Capital (£)
Airspeed	(L) Fokker, Douglas aircraft (P) Retractable undercarriage	160,000
General Aircraft	(P) Monospar wing system	375,000
Phillips and Powis	(D) Various aircraft	125,000
British Marine	(L) Sikorsky flying boat	400,000
Parnall	(D) Fraser Nash gun turret	300,000
Aero Engines	(L) Hispano Suiza engine	275,000
Pobjoy	(L) Short Scion aircraft	275,000
ACGB	(L) Aeronca training aircraft	200,000
Aircraft Components	(D) Aircraft equipment	200,000
Total		2,310,000

Source: *The Statist.*

The aircraft boom also created opportunities for companies normally interested in other sectors of the engineering industry. The most successful was the English Electric Company. English Electric, which had designed and produced aircraft during the Great War, was initially asked by the Air Ministry to produce just 75 Handley Page Hampden bombers, the intention being to 'train a labour force and set up an organisation for aircraft construction which would be available in an emergency'.[21] The company was then invited by Freeman to assume further responsibilities.

The managing director initially responded with some caution. He was, he maintained, under pressure from the Board of Trade to maintain exports. Moreover, concentration on munitions work during the First World War had led English Electric to neglect the civil side of its business and this had caused severe trading problems during the 1920s. He argued that civil business might again be jeopardised if English Electric took on more aircraft work, and he therefore asked the Air Ministry 'to give him such assistance as was within their power to enable him to build up an aircraft design section, and . . . not to withhold orders from him after the war if he was able to form such a section'. He felt that English Electric would influence design so that aircraft could be produced more economically, and promised to produce the Hampden more cheaply than Handley Page.[22]

The documents do not record Freeman's reaction. However, as he was Chief Executive of the Ministry of Aircraft Production when English Electric bought out Napier at the end of 1942, and when the company acquired Westland's chief designer, Edward Petter, together with an embryo bomber design (which became the highly successful Canberra) in 1944, it seems unlikely that he discouraged the firm from re-entering the design field.[23]

A number of shipbuilders also became involved in aircraft production through buy-outs and joint management deals. By 1938, Swan Hunter & Wigham Richardson, John Brown and, of course, Vickers, all owned aircraft companies. Denny was associated with Blackburn, and Harland & Wolff with Short Brothers.[24] The shipbuilders provided a much-needed injection of capital into some of the smaller aircraft firms. Westland's financial position remained somewhat unsound until John Brown acquired a controlling interest in the company in 1938,[25] and Airspeed's finances were equally parlous until the company's association with Swan Hunter.

Airspeed was formed in 1931 with just £5000 of private capital.

The managing director, Nevil Shute Norway, later recorded that, in its first three years, the company was engaged in an almost continuous struggle to contain its overdraft and to satisfy its creditors, and had only continued trading thanks to the willingness of its principal shareholder to provide more capital and to make personal guarantees to its bankers. However, by the beginning of 1934, Norway recalled:

the finances of the company had grown to such a scale as to make it impossible for our early loyal supporters to carry the company much further. In our continuing quest for fresh capital we made contact with a City firm who seemed to think they could place a public issue for us . . . [but] our continuing losses [were] something of an obstacle to a successful issue.[26]

The aircraft boom transformed this situation. Airspeed began 'to attract the attention of more conservative City houses. One . . . was closely linked with the well-known shipbuilders Swan Hunter and Wigham Richardson Ltd' which was contemplating taking an interest in aircraft manufacturing.

By the beginning of June [1934] an agreement had been reached . . . for a public issue of shares in Airspeed Ltd under the auspices of Swan Hunter, who would acquire control of the company through an ordinary shareholding . . . After three years substantial capital was at last in sight.[27]

Swan Hunter's chief interest was, in fact, civil, rather than military, aviation; the company envisaged using its shipbuilding facilities to produce flying boats. Yet these hopes were never fulfilled. Shipyards which had been empty in the early 1930s were soon employed on Admiralty contracts for rearmament and in the meantime Airspeed's civil 'Envoy' was converted into the Oxford, which became the most successful RAF trainer of the Second World War.[28] Despite the Oxford's success, Swan Hunter did not welcome the outbreak of hostilities in 1939. It was anticipated that Oxford production would come to dominate Airspeed's activities to the detriment of design work for the future, that there would be a glut of second-hand aircraft on the world market in the immediate post-war years and that, as a result, it would be very difficult to win either civil or military contracts. So in May 1940, only two years after Airspeed began to trade profitably, Swan Hunter sold its stake in the company to De Havilland.[29]

The motor car industry might perhaps have adopted a similar approach to the shipbuilders had it not been for its involvement in the shadow scheme. Two firms, Alvis and Wolseley, did attempt to establish themselves as independent contractors to the Air Ministry, but both failed completely. Alvis incurred substantial losses, while the Nuffield organisation eventually closed down Wolseley's aero-engine division altogether.[30]

The aircraft boom peaked in 1936, when the industry's demand for new capital was at its height, but as firms issued more capital share prices naturally tended to fall. In July 1936 ordinary £1 shares in De Havilland were valued at 73s. but by January 1938 the price had fallen to 40s. Rolls-Royce shares fell from 171s. to 96s. between August 1936 and June 1938; Hawker Siddeley's fell from 39s. to 29s. between October 1936 and April 1938.[31] A similar pattern prevailed throughout the industry. After 1936, firms often succeeded in fulfilling their financial requirements by capitalising reserves and by borrowing, and the need to float new issues was, in any case, reduced after the government began to finance fixed capital extensions in 1938. Nevertheless, further cash issues, like the £2 million Hawker Siddeley issue of November 1937, were still heavily over-subscribed.[32]

By the outbreak of war, then, the share capital of the principal aircraft firms had been raised to more than £19 million.[33] In the words of the Official Historian, 'the figures indicate that the pre-war aircraft industry itself financed much of its expansion'.[34] In 1935 alone, aircraft and aero-engine share issues accounted for 9.5 per cent of total new issues in Britain's commercial and industrial sector.[35] The extensive employment of subcontractors after 1938 ensured that private capital continued to play an important role in financing aircraft production throughout rearmament and well into the war. In 1941 a total capital of £45.9 million was employed in manufacturing the products of eight leading aircraft firms. Of this sum, 39 per cent was supplied by the professional industry, 48 per cent by subcontractors, and just 13 per cent by the state (see Table 32).

THE SUPPLY OF WORKING CAPITAL

Although the government relied heavily on private capital to finance aircraft production, overcapitalisation would have benefited neither producer nor consumer. The Air Ministry's view on this subject was clearly expressed by the Secretary on the occasion of Hawker's flotation in 1933: 'Their capitalisation is beyond doubt too heavy',

Bullock wrote, 'and we, as their principal customers, will inevitably be asked excessive prices for their products to enable them to pay dividends on this inflated capital.'[36]

TABLE 32

Private, Subcontract and State Capital (£000) Employed in Aircraft Production, 1940–41

Firm	Company's Capital	Subcontract Capital	State Capital
Blackburn	2,400	2,013	619
Boulton Paul	1,400	964	46
Bristol	6,925	5,987	2,055
De Havilland	1,700	1,741	41
Handley Page	1,440	1,133	259
Rolls-Royce	3,079	8,173	2,370
Short	1,600	1,367	450
Westland	864	625	73
Total	18,008	22,003	5,913

Source: PRO AVIA 10/253.

In fact, Bullock was wrong. Although Hawker ultimately required a share capital of £6 million to fund the amalgamations of the following three years, this was by no means excessive considering the number of different subsidiaries involved and, after the creation of the Hawker Siddeley combine had been accomplished, the holding company came to rely for finance on £5 million worth of short-term debentures. The advantage of this method lay in the fact that when the finance was no longer needed the debentures could be repaid so that the group was not burdened with capital which it did not require.[37] The general view of the industry was more accurately summarised by Frederick Handley Page in 1939: 'Our endeavour', he wrote, 'is to turn over our capital as many times as possible during the year and achieve a high rate of production with a relatively small capital.'[38]

The provision of state capital assistance was therefore in the mutual interests of the government and the contractors, and this had been acknowledged throughout the inter-war years by the provision of staged 'progress payments' covering up to 80 per cent of the estimated cost of completed work. At the beginning of scheme C, the Air Ministry, under pressure from the industrialists, agreed to make monthly payments of 80 per cent of the money spent on the purchase of materials and to raise the limit of progress payments for materials to 90 per cent of their estimated total cost.[39]

Liberal as these advances were, they often seemed insufficient because of the sheer scope of the expansion programme. According to a report by the Air Ministry's Director of Contracts

[p]ayments made on the basis of a provisional price related in the main to completion of stated stages of work, are inadequate, partly because the Department must necessarily err on the side of caution in arriving at the provisional price and partly because heavy expenditure is incurred before completed stages of work are finished. No advances are made against tool expenditure. No convenient system exists for making progress payments against spares contracts . . . The liquid assets normally held by the firms have been largely expended on capital items.[40]

Pre-production work for scheme F imposed greater financial burdens on the industry than it had borne at any time in its history, and the technical problems encountered in developing the new generation of military aircraft resulted in a tremendous accumulation of expenditure before deliveries started. By the time Handley Page delivered the first Harrow bombers, the company's financial resources had been severely depleted and it had to borrow to fulfil requirements. The cash position improved as production accelerated but deteriorated again during the transition from the Harrow to the Hampden, when Handley Page had, once again, to rely on bank loans and an overdraft.[41]

The pattern was typical. More than 40 per cent of total contract expenditure for both the Wellesley and the Wellington bombers had been incurred by Vickers Aviation before the first production aircraft were delivered, and between July 1936 and December 1937 the Vickers parent company advanced more than £800,000 to its aircraft subsidiaries.[42] Fairey's 1936 share issue provided £625,000, but considerably more than this amount was eventually required to finance the improvement and extension of capacity and the expansion of turnover. Fairey's accounts show that its Stockport subsidiary, where the problematic Battle was being developed, became heavily indebted to the parent company. By January 1937, Fairey's cash resources had been exhausted, outside investments had been sold and an overdraft had been raised (see Table 33).[43]

For some companies the search for additional finance must have seemed interminable. Blackburn was floated on the Stock Market in 1936 with an issued capital of £630,000. In June 1937, the company's bankers granted overdraft facilities of £300,000. Additional requirements for fixed and working capital led Blackburn to increase its issued capital to £930,000 in September 1937, and in December overdraft facilities were doubled. They were increased again to £700,000 in June 1938 and to £850,000 in October.[44] These sums were required in addition to progress payments from the Air Ministry which totalled £600,000 prior to the delivery of one type and £1.8 million before the emergence of another.[45] The demand for finance was in large part due to the technical problems which Blackburn encountered in the development of their designs.

TABLE 33

Fairey Aviation, Financial Position, 1935–36 (£)

	1935	1936
Share Capital	500,000	1,000,000
Overdraft	—	60,000
Creditors	91,998	367,044
Subsidiaries a/c	31,217	729,184
Investments	11,7238	—
Cash	75,727	943

Source: *The Statist*, 18 December 1937.

It was not always easy for smaller firms to obtain the necessary funds. Peter Acland of Westland wrote to Frederick Handley Page in 1937 that

Westland got caught at the beginning of the expansion period with no work, and even to date have not had an even flow of production, which renders it reasonable to expect a return to be earned for shareholders . . . There are considerations of obtaining further capital to execute an agreed output, and there is extraordinarily little in the way of a tale to tell to the financier.[46]

The position improved somewhat when the progress payment system was extended in 1937 to cover jigs and tools but the Air Ministry was otherwise reluctant to make concessions. It was usually argued that the retention of a portion of the production costs served as an incentive to greater output. There was naturally a reluctance to allow contractors to profit from 'a mass of work at an uncompleted stage' or to work on what amounted to a 'cost-plus' basis, but there is also little doubt that by retaining 10 or 20 per cent of a very conservative provisional price, sometimes amounting to only 50 per cent of actual costs, the Ministry was seeking to expedite contractual negotiations with the aircraft firms.[47] As the balance between progress payments and actual costs had to be financed by the contractors, and as the Air Ministry did not allow interest on loans as an overhead, it was to the firms' own advantage that final prices should be fixed as quickly as possible.

There were therefore no further changes to the progress payment system until, in the later months of 1938, the acceleration of rearmament began to create financial problems for the industry. A new progress payment stage covering detail parts was introduced but this did not prevent two of the three parent firms of the new heavy bomber groups, Handley Page and Short, from protesting about the inadequacy of existing provisions for state capital assistance. Both companies operated on a relatively small scale in the inter-war years and lacked the necessary resources to finance pre-production work on the scale which the group schemes entailed. The Stirling group, for example, required £7 million worth of jigs, tools and materials. Had the Air Ministry retained just 10 per cent of this sum Short Brothers would still have had to finance the outstanding £700,000.[48]

There were also complaints about the Air Ministry's requirement that contractors should produce paid bills in support of their claims for reimbursement. This stipulation, reasonable enough under normal circumstances, now resulted in firms like Handley Page having to finance very substantial payments for materials for several weeks. Under pressure from his own company secretary, Frederick Handley Page suggested to the Air Ministry that firms might withhold payment for materials pending receipt of each monthly payment from the Air Ministry. When the Ministry refused, Handley Page abruptly withdrew his financial support for the London Aircraft Production group of Halifax bomber subcontractors. An alternative approach was adopted by Hawker Siddeley which, in 1940, furnished the MAP's Directorate of Contracts with false certificates of payment.[49]

A general dissatisfaction with the progress payment system developed following the outbreak of war. In November 1939 the managing director of Vickers-Armstrong urged his aircraft division to reduce its indebtedness: 'The Aviation Department is getting thoroughly unpopular here', he wrote, 'owing to the tremendous amount of financing we are called upon to provide. I hope every effort will be made to put the company in the position to claim these instalments [from the Air Ministry].'[50] Sir Frank Spriggs complained that Hawker Siddeley was having to finance £2 million because of delays in settling prices but, by November 1940, Bristol was owed more than twice this sum by the government (see Table 34). 'Apart from any increased production programmes', the company secretary wrote, 'we have undertaken . . . additional work concerned with salvage, the repair of aircraft and engines, the repair of components and accessories, [and] the supply of large quantities of spare parts.'[51]

Expenditure had also been incurred because of dispersal, air raid precautions and bomb damage, and enemy action had interrupted production, causing an increase in overheads which was not expected when provisional prices were agreed. Some assistance had been provided by the company's bankers but it was now felt that the volume of work involved 'the provision of an excessive amount of working capital . . . so that even with the most prompt method of payment which was practicable, the amount due to the Company at any one time was very considerable'.[52]

TABLE 34

Amounts Owed by the Ministry of Aircraft Production to the Bristol Aeroplane Company, 31 October 1940

Division	Amounts Owed by the MAP (£)
Aircraft	1,232,850
Engines	1,990,994
Other work	1,658,632
Claims	342,040
Less amounts due to the MAP	−939,440
Total	4,285,076

Source: PRO AVIA 15/3710.

It was, in fact, too easy for the industry to blame the government for all its difficulties. Privately, the firms recognised that the limitations of state capital assistance were only part of the problem; financial requirements could be reduced by improved production methods and stock control systems,[53] but given the rapid increase in turnover which followed the outbreak of war it was inevitable that the state's role in providing working capital should have been extended. New capacity sanctioned at this time was invariably managed on an agency basis, similar to that employed in the shadow factories, whereby companies received a management fee and the

state underwrote all production expenses. All 12,000 Spitfires produced at Castle Bromwich, and most of the 12,000 Wellingtons
built by Vickers-Armstrong were, for example, subject to agency
contracts.

However, commercial operations continued on a large scale and it
therefore became necessary for the government to liberalise the
progress payment system. In April 1940 the Treasury agreed to
allow progress payments of up to 95 per cent, and in Short Brothers'
case 97½ per cent was allowed for materials, jigs and tools. Handley
Page's representations resulted in progress payments being made
for spare parts. Further concessions were granted during the
emergency of 1940, and when companies still found themselves in
financial difficulties (as in the case of Bristol) special advances were
made by the MAP which, in practice, often amounted to at least
£1 million.[54]

The government also reached an agreement with the Bank of
England and the clearing banks under the terms of which overdraft
facilities were extended and normal requirements for securities on
loans were waived. In the words of the Official Historian:

The object was to ensure that the banks would readily meet the requirements of Government contractors and subcontractors for working capital
and that the difficulties of finance would cause no impediment to the flow
of production.[55]

These measures were mostly successful and the industry was able
to fund its operations by borrowing on a far greater scale than was
normal in peacetime. In July 1940, for example, De Havilland's
company secretary reported to the board that its bankers 'had
agreed to provide overdraft facilities up to a limit of £750,000'. As
this exceeded the borrowing powers conferred by De Havilland's
articles of association, a meeting of shareholders was called and a
borrowing limit equivalent to twice the company's issued capital
was authorised.[56]

There were two other ways in which the firms could finance
the growing volume of production. The first was to take the maximum possible advantage of credit from material and component
suppliers. Credit provided the most important source of private
finance for the aircraft industry in the rearmament years,[57] and the
more liberal stance adopted by the banks in 1940 had the effect of
adding to the amount of 'trade' credit which contractors could
advance to one another. Requirements for trade credit did not,

however, grow in proportion to the industry's turnover after the outbreak of war. This was in part because materials, equipment and components were increasingly purchased directly by the government's supply departments and issued free to main contractors.[58] But at the same time firms began to employ more of their profits as working capital. General reserves were built up, but more important were taxation reserves, which enabled taxable earnings to be used to finance production during the nine to eighteen months between the receipt of those earnings and the date on which payment to the Inland Revenue became due.[59]

Of course, this effectively meant that the state was providing another form of finance to industry in addition to the progress payment scheme. The industry's dependence on taxation reserves was denounced by the Ministry of Aircraft Production as 'an illegitimate form of borrowing', and it was doubtless for this reason that the MAP continued to reduce the profit allowances paid to the firms and to monitor their costs closely even after the introduction of Excess Profits Tax (EPT) in 1940.[60] If the industry had been prevented from using its profits in this way, however, it would only have demanded more direct assistance from the government, and the latter, as already noted, preferred to set clearly defined limits on the proportion of production costs for which progress payments could be claimed. It was preferable to allow industry as much financial independence as possible and in 1942 the MAP even considered asking the Inland Revenue to defer its demands for EPT.[61]

The finance provided by bank overdrafts, trade credit and taxation reserves was very substantial indeed. In July 1941, for example, the private capital employed by the five Hawker Siddeley firms amounted to £14.3 million.[62] In order to execute a turnover of £42 million, additional finance amounting to no less than £14.5 million was required, divided between bank overdrafts (£3.5 million), sums owed to subcontractors and suppliers (£6 million), and amounts due to the Inland Revenue (£5 million).[63] In the same year, Handley Page, which employed a private capital of £1.5 million, halved the interim dividend on its ordinary stock and raised its taxation reserve from £380,532 to £745,000. The company owed an equivalent sum to its trade creditors (see Table 35). In short, overdrafts, trade and taxation credit provided the aircraft firms with finance equal in value to their own capital.

Industry and air power

TABLE 35

Handley Page Liabilities, Trade Creditors and Reserves (£), 1938–41

Year	A) Trade Creditors	B) Taxation Reserve	C) General Reserve	D) Total Liabilities	% A+B+C to D
1938	273,987	164,495	80,000	1,383,187	37.5
1939	416,231	414,726	160,000	1,791,522	55.3
1940	412,162	380,532	200,000	1,969,679	50.4
1941	751,893	745,000	230,000	2,535,873	68.1

Source: Handley Page archive, Handley Page annual accounts. Total liabilities do not include amounts received on account against certain contracts.

THE MCLINTOCK AGREEMENT

Government financial aid to the aircraft industry was progressively liberalised between 1935 and 1941. In contrast, state control over aircraft prices and profits became increasingly strict as the volume of output increased. Contractual policy was intended by the government to perform three principal functions. First, public pledges to prevent profiteering had to be honoured. Second, it was felt that prices should provide an 'incentive to production and efficiency'.[64] Third, the aircraft industry had to be given clear assurances about the level of remuneration which it could expect. As Lord Weir pointed out, it is 'difficult to secure the best enthusiasm and energy on the part of a contractor until he knows he can go ahead with confidence in being fairly treated financially'.[65]

In 1935 it appeared to the National Government that there were overwhelming economic and political objections to imposing statutory controls on firms engaged in rearmament work, and consistent attempts were therefore made to negotiate a general financial settlement with the aircraft industry. This settlement came to be known as the McLintock Agreement after the prominent accountant who negotiated on behalf of the SBAC. The McLintock Agreement was only concluded after months of bargaining and proved extremely difficult to implement effectively. It was revised at the end of 1937, completely renegotiated in 1939, and finally abandoned altogether in 1941. From an industrial standpoint the interests of the government and the aircraft firms were broadly compatible: the government needed to buy aircraft for the defence programme; the industry wanted to produce and sell aircraft for a reasonable return. The political controversy generated by rearmament profits, however, tended to obscure this community of interests. The government's three contractual objectives proved irreconcilable until compulsory profit controls were imposed on the outbreak of war in 1939.

Historians such as R.P. Shay have argued that the aircraft firms were largely responsible for the failure of contractual policy in the rearmament years. The evidence for this view is supplied, predictably enough, by the files of the responsible government departments. It is, however, now also possible to consult the files of Frederick Handley Page, a former chairman of the SBAC, and those of leading aircraft contractors such as Vickers and Hawker-Siddeley. The records demonstrate that while the industrialists were naturally eager to maximise their profits, they were not as unreasonable as has sometimes been supposed.

The government's first public assurances that excessive profits would not be permitted on rearmament contracts left the aircraft industry in an uncertain financial position in the summer of 1935.[66] If profits were to be controlled, how large a return could be expected? Substantial commitments were being made. Bristol Aeroplane and Hawker Siddeley were floated on the Stock Market. Vickers, Supermarine, Fairey and Handley Page each initiated schemes for expanding factories and plant. The value of De Havilland's fixed assets increased by more than £100,000 between 1934 and 1935.[67] The industry considered that much of this capacity would be excess to normal requirements and the government was therefore asked to provide compensation for plant which became redundant after the completion of scheme C in 1937.[68] There was considerable anxiety when assurances on this point were not immediately forthcoming.

Another contractual problem was raised by the inadequacy of the government's costing machinery. The Air Ministry had controlled prices during the early 1930s by inviting firms to compete for available contracts. The Ministry also made its own 'technical' cost investigations but only limited importance was attached to their accuracy because orders were generally small.[69] Rearmament changed this situation completely. As demand began to outstrip supply, the scope for competitive tendering declined and the growing volume of turnover enlarged both the absolute amount of profit earned by the industry and the margin by which actual production costs undercut the Air Ministry's cost estimates.[70] The Ministry's approach to fixing prices consequently became increasingly circumspect.

By the later months of 1935, price fixing and other financial questions were causing so much concern to both the Air Ministry and the aircraft industry that the two sides sought agreement on a general set of contractual principles covering rearmament orders. The negotiations began inauspiciously. The SBAC (representing the industry) wanted disagreements on contractual questions to be settled by independent arbitration whereas the Air Ministry maintained that ultimate control over such questions rested with the Secretary of State. The industry was also concerned about the attempts of new aircraft companies like General Aircraft to bid for Air Ministry contracts, and about the Ministry's refusal to fix prices for established designs. On the question of profits, the SBAC 'did not think it was practical to ask for a fixed rate . . . but it was necessary that for every contract a definite rate should be fixed'.[71] Until such issues were resolved there was a danger that the industry would delay the implementation of further expansion plans.

There followed several months of inconclusive bargaining during which scheme C was succeeded by the much more ambitious scheme F. The question of compensation for capital which became redundant after the completion of the expansion scheme (the so-called 'Capital Clause' in the prospective agreement) became a major bone of contention, as did the industry's insistence that development expenditure was a legitimate overhead charge. Frederick Handley Page wrote that

[u]ntil this is arranged I very much fear that our profit may gradually disappear into the background and finally be represented by bricks and mortar and an intangible value in patents and designs. We shall have to press very hard in order to get a decision, and the most likely procedure on the part of the Air Ministry is to postpone things until the end of the expansion programme.[72]

Yet the Air Ministry was initially unsympathetic to the industry's arguments about research and development expenditure and indicated that Capital Clause compensation would be withheld if contractors made 'excessive' profits. Handley Page told McLean that

[t]his interlinking of capital expenditure with profit immediately makes any settlement of price provisional in that if by efficiency of manufacture, cleverness in design, or by any other means which a capable manufacturer

may take, a large profit is made on the contract, then there will be a very difficult case to make out for a claim against the Minister.

Nor was there any agreement over the price of established types of aircraft. McLean pointed out that such designs would account for no more than 10 per cent of scheme F contracts, and 'suggested . . . that [the Air Ministry] should agree to accept the technical cost investigations already made and, by adjusting prices to the current levels of overheads, agree a final price'. The Air Ministry, however, maintained that it lacked adequate data owing to a shortage of costing staff.[73]

By March 1936 an impasse had been reached. 'We have been put in a position that is quite unique', McLean recorded. 'Without process of or reference to Parliament and without Statutory Powers of any kind, the Air Ministry [has] commandeered our assets and pledged our credit up to the hilt without offering us any consideration that we could consider "valuable".' It was in an attempt to overcome this deadlock that the industry sought the advice of Sir William McLintock. According to Charles Allen, the SBAC's solicitor, McLintock rapidly acquired 'a very clear grasp of the situation', and the society subsequently invited him to negotiate on its behalf.[74]

McLintock clearly provided a significant stimulus to the negotiations. On 20 May 1936, Swinton agreed to treat development spending as an overhead charge and extended the Capital Clause for the duration of scheme F.[75] For its part, the SBAC reluctantly accepted the reservation regarding 'excessive' profit which was insisted on by the Cabinet.[76] It is unclear whether expansion was significantly delayed by the industry's insistence on compensation for redundant capacity. Some firms effectively committed themselves to expansion before May 1936. At least one other, Blackburn, decided as late as 18 May to 'keep capital expenditure . . . to the minimum . . . We were uncertain to what extent the government would enable us to recover losses on capital expenditure.'

This element of uncertainty soon disappeared and by the end of 1936 the industry had spent £3.8 million on factory extensions which were expected to be covered by the Capital Clause. Blackburn alone had spent £550,000 by September 1937.[77]

By July the outstanding contractual issues had been narrowed down to profit rates and pricing methods. Profit rates were in practice negotiated on a contract-by-contract basis and varied from 7 to

10 per cent until 1938, when the Air Ministry reduced the average allowance to 7 per cent. Where pricing formulae were concerned there were two obvious solutions. One was the fixed price, which had been employed for much of the inter-war period and which was favoured by industrialists like Frederick Handley Page.[78] Handley Page had, by this time, developed so-called 'split' construction and was presumably fairly certain that production costs could be substantially reduced.[79] Other contractors wanted the second solution – provision for 'cost-plus' when no agreement on fixed prices could be reached.[80] Their difficulty lay in the technical revolution which was then taking place in aircraft design. The industry was, in its own words, engaged in the 'construction of a fundamentally different article' which involved radical changes in manufacturing methods and consequently in production costs. The government was, however, vehemently opposed to the general application of 'cost-plus' (or 'time-and-line', as it was known) because it had been associated with profiteering during the First World War and because it actually gave contractors an incentive to inflate their costs.[81] It was felt that costing should be confined to small batches of aircraft and used purely as a basis for the negotiation of fixed prices for the bulk of the contract.

The resulting deadlock led to a protracted search for a third price formula which was neither fixed price nor time and line. Sir Robert McLean of Vickers eventually proposed a 'maximum price' scheme whereby the actual costs of early production aircraft would be used to agree the price for the remainder of the contract. If costs fell below the maximum price the savings would be shared by the Air Ministry and the contractor, but if the price was exceeded the loss would be borne by the contractor alone.[82]

Yet the inclusion of this third formula in the final draft of the McLintock Agreement was chiefly a matter of politics rather than practicalities. The firms felt that they could not reasonably be expected to bear any losses on contracts for new types, and the agreement, which was drawn up in September, allowed production to commence on a time-and-line basis pending the settlement of a fixed or maximum price. When the SBAC accepted this provision, Swinton's immediate reaction was to do likewise, although he insisted that he would not accept time-and-line if costs had been deliberately inflated, and replaced the emotive phrase, 'time-and-line', with 'ascertained cost plus a profit'.[83] McLintock accepted these terms on the SBAC's behalf on 8 October.[84]

THE OPERATION OF THE FIRST MCLINTOCK AGREEMENT

Any lasting contractual settlement had to benefit both state and industry. The first McLintock Agreement benefited neither. 'It is becoming increasingly clear that the industry desire to drift indefinitely so that they may secure in effect time-and-line contracts', the Air Ministry's Deputy Under-Secretary, Sir Henry Self, wrote in October 1937. Apart from the political objections to this situation, there was also the danger that contractors might inflate their costs. Yet the delays in price fixing were also caused by the Air Ministry's own costs and accounting procedures. There was, in Self's words, a 'natural desire' among the technical cost staff 'to delay the estimates until all practical danger of error is eliminated'.[85]

Production problems associated with new types of aircraft exacerbated this 'danger of error'. The Air Ministry's Principal Technical Cost Officer reported, for example, that it had been impossible to estimate labour costs on the Armstrong Whitworth Whitley bomber because the firm 'put in temporary piecework prices . . . which [only] covered the first period of production'. The prices had been modified when tools became available and were modified again when the tools were improved. In the case of the Bristol Blenheim it was pointed out that there had been numerous alterations to the design:

It was not until a very long way through this contract that they even settled the wing design. We have, in fact, made two estimates on the greater portion of this machine because of the modifications to the design.

Of another type, the Vickers Wellesley, it was stated that the geodetic design principle was so novel that actual costs were to be paid for the first 96 aircraft.[86]

If this was unsatisfactory from the government's viewpoint, it proved equally objectionable to the industry. The firms preferred to fix prices before the commencement of series production if at all possible because any subsequent fall in costs then increased their profit margins. Indeed, despite the Air Ministry's anxiety about the excessive use of time-and-line, figures compiled by the SBAC indicate that fixed prices were agreed for the majority of turnover in both 1937 and 1938.[87]

The advantage for the industry is illustrated by the cost records for 500 Hampden bombers produced by Handley Page between 1938 and 1940. In a fixed price deal negotiated at the beginning of production, the Air Ministry overestimated the cost of the first 180 aircraft by 2 per cent. The Air Ministry reduced its estimate for the remaining 320 aircraft but, as Handley Page also lowered production costs, the 2 per cent margin was retained. Handley Page earned an extra £100,000, which was unacceptable from the political standpoint, but the state did benefit from the higher output and more efficient use of resources.[88] The fixed price thus provided some scope for a *modus vivendi* between state and industry whereby improvements in efficiency benefited both parties.

In contrast few gains were derived from time-and-line contracts. Supermarine's contracts manager wrote that

I am of the opinion that 'batch costing' is to be avoided as much as possible, as it discloses all sectional and component costs of the machine, and invariably means that a lower standard of profit is admitted for both machines and spares.[89]

Vickers was dismayed by the profit levels allowed on costed contracts. 'We do not accept the view that in regard to the first [Wellesley] contract a rate of profit of 7½ per cent on our work is adequate remuneration', the company complained to the Air Ministry. 'The rate of profit proposed by your Department is not a fair one and we are only accepting it under protest.'[90]

These lower rates seemed reasonable to the Air Ministry because of the elimination of commercial risk to the contractor, but the firms felt that they were paying twice for the added security because the technical problems which led to the employment of the time-and-line formula, with a lower profit allowance, also resulted in a reduced volume of turnover and, correspondingly, of profit during the financial year.[91] Between 1936 and 1937, when Rolls-Royce was introducing the Merlin, net aero sales fell by 4.4 per cent and gross aero profits by 25 per cent, while share capital was increased by 29 per cent.[92] At Hawker, delays with the Hurricane caused turnover to fall from £1 million in 1936 to £730,000 in 1938 while net profits fell by 23 per cent in the same period.[93] In both 1935 and 1936, Vickers Aviation did not receive any dividend from Supermarine, and problems with the Wellesley resulted in a decline in the net profits of Vickers Aviation of 24 per cent between 1936 and 1937 despite substantial cash advances from the holding company.[94] The average percentage of disclosed net profits to total capital and reserves in the aircraft industry only increased from 13.6 to 16.8 between 1935 and 1938.[95]

The failure of the first McLintock Agreement must be viewed in this context. In 1935 the SBAC had assured the Air Ministry that there would be no profiteering from rearmament contracts, doubtless anticipating that the increased volume of turnover would compensate for any percentage reduction in the customary profit allowance. Yet turnover failed to expand at anything like the expected rate and much of the capital invested in the industry in this period did not fulfil its true earning potential. This is illustrated by the fact that individual contractors like Handley Page, which overcame their production problems, generated a much higher turnover than the remainder of the industry. According to a survey of 15 aircraft firms (covering non-government and government work) the average ratio of capital to turnover in 1937 was 1:1.42 but the optimum ratio was 1:3. The average rate of profit on capital amounted to 18 per cent (a rate which proved irreducible even after the imposition of wartime controls) but the rate of profit on 6 per cent of the capital was again more than double the average rate.[96] Had it not been for the technical problems encountered by the industry in 1936 and 1937 it is clear that this average would have been very much higher.

At the end of 1937 Sir Henry Self brought sufficient pressure to bear on both the firms and his own costing staff to accelerate negotiations, and by April 1938 prices had been agreed for several new types including the Blenheim and the Battle.[97] Actual costs were paid for more aircraft than the government would have wished during the early stages of rearmament, but the firms were generally eager to agree fixed prices after their initial production problems had been overcome.[98] However, the prolonged efforts to negotiate a third pricing formula were largely wasted. Maximum prices were rarely employed because of the industry's refusal to accept any risk of loss on new types of aircraft and its aversion to sharing any savings with the Air Ministry.

In the meantime, the Treasury Solicitor raised a number of legal objections to the McLintock Agreement, and the Air Ministry was told by the Public Accounts Committee that the terms should be revised so that they could be applied individually to the firms concerned. As the SBAC thought that an agreement had effectively been finalised in 1936, it was initially baffled by this recommendation: 'I do not understand why either the Air Ministry or the Public Accounts Committee should think it necessary to have a still further document', Allen wrote to Handley Page.

The terms settled between McLintock and the Air Ministry are recorded in correspondence and are summarised in Sir William's report to the Council of the SBAC . . . The SBAC wrote to the Air Ministry formally recording the Agreement entered into on its behalf by McLintock; and the SBAC letter had attached to it a list of Ordinary Members of the Society on whose behalf Sir William McLintock was authorised to negotiate. I therefore looked upon this phase of the matter as concluded and that no further document was necessary.[99]

The industrialists were understandably concerned that an agreement which was already in many respects unsatisfactory to them was apparently to be altered in the Air Ministry's favour. Their fears proved justified, for while the Ministry now attempted to make the third pricing formula (renamed the 'basic price') more attractive to the industry, it demanded one fundamental concession in return – the renunciation of the time-and-line clause in the original agreement. This was eventually achieved through the inclusion of a new clause providing for arbitration when neither a fixed nor a basic price could be agreed.[100]

The SBAC's initial opposition to the revised Air Ministry proposals must be viewed in the context of the relatively limited scope of British rearmament before the spring of 1938. Until then there was no certainty that production would be maintained after the completion of scheme F; on the contrary, it seemed far more likely that RAF orders would be drastically reduced. In the circumstances the firms were, reasonably enough, determined to profit from what might well have been only a short-term increase in demand, and the time-and-line clause was therefore surrendered with considerable reluctance.

Indeed, the revised terms were only agreed after the SBAC received informal notification that a new defence programme was shortly to be initiated. For those firms that had largely been engaged in design and development work since 1935 there was now the attractive prospect of straight-run production at fixed prices. Alternative methods of pricing would be unnecessary and Air Ministry contracts would become much more profitable. Hence, while the revised McLintock Agreement initially seemed more favourable to the state from the political standpoint than its predecessor, in reality it was excessively generous to the industry. 'The more I look at the McLintock arrangements', Allen wrote to Handley Page after the *Anschluss*, 'the more convinced I am that McLintock really did his very best for us – and I think he did it very well.'[101]

THE SECOND McLINTOCK AGREEMENT

Between 1936 and 1938 lengthy discussions had taken place in government circles about the way in which the profit rate on aircraft contracts should be calculated. The Air Ministry considered that profits would lose their incentive value unless computed as a percentage of the cost of each contract, and this argument had prevailed. The Treasury, on the other hand, consistently maintained that the industry's earnings could be more effectively limited by calculating profit as a yearly percentage of the capital employed by the contractor. This was superficially correct but it was also a charter for inefficiency, as Frederick Handley Page pointed out in 1939:

If a firm, because of efficient working in the past and prudent methods of finance, has been able to build up a large business with a small capital, they should not be penalised and allowed because of their smaller capital a lesser rate of profit than a firm whose inefficient methods and extravagant policy have led them to the employment of a large volume of capital.[102]

The logic of this contention was unanswerable. In 1937, for example, Handley Page assembled more bomber fuselages with one set of jigs than Armstrong Whitworth assembled with 16 sets,[103] yet under the Treasury formula Armstrong Whitworth would have earned a larger profit. This was, however, a difficult argument to employ against political criticism of aircraft profits. The Treasury's concern was that the statutory requirements of the period compelled companies to publish in their annual reports dividends, profits, and nominal capital, but not sales. Charges of profiteering would therefore be based on the relationship between profits and nominal capital, even when contractors were earning a relatively small profit on turnover.[104] The problem was exacerbated by the fact that the total capital of most firms greatly exceeded their nominal capital. It did not necessarily follow that a high dividend rate on issued capital connoted an excessive profit but it was easy for Labour MPs to claim that it did.[105]

As already noted, technical constraints prevented much of the capital employed by the aircraft industry from fulfilling its true earning potential between 1935 and 1938, but by the later months of 1938 most firms had overcome their production problems. Capital was being employed much more efficiently and was realising profits on a scale which caused controversy in political circles. This did not, in itself, provide grounds for terminating the McLintock

Agreement. Rather, justification was provided by the government's decision, in 1938, to invest directly in buildings and plant for aircraft production. Very little of this capacity would have become productive before 1939, but it would subsequently have allowed the industry to earn an even higher return on its own capital.[106] In February 1939 a specially appointed Treasury committee therefore recommended that the McLintock Agreement should be revised, and on 1 March Swinton's successor at the Air Ministry, Sir Kingsley Wood, duly advised the SBAC that the government intended to negotiate a new agreement.[107]

Official investigations into aircraft industry profits had revealed that the highest earnings were obtained from fixed price contracts.[108] For the third time, therefore, the Air Ministry attempted to devise an alternative pricing formula.

The industry was now advised that fixed prices would be negotiated only in exceptional circumstances. Instead, actual costs would be paid for initial batches of aircraft, after which 'target' prices would be used, with a fixed profit allowance determined mainly by the amount of private and government capital employed. The maximum rate of profit on turnover would be 5 per cent. If costs fell below the target price the savings would again be shared between the government and the firms but, if the target price was exceeded, the contractor would bear the entire loss.[109]

The industry's response may best be summarised by Handley Page's company secretary:

On initial batches of a new type contract the maximum of 5 per cent [profit] is not equitable. The change over of a new contract usually involves a partial break in the flow of production, and during such a period a higher ratio of profit should be allowed as compensation . . . Target costs will be difficult to agree . . . for the reason that cost data of the previous batches will be available, and this new information, together with estimates from [the Air Ministry's] Cost Estimating department, will place [its] Contracts Department in a more favourable position when negotiating . . . The result may well be entire disagreement or minimum profit.[110]

The industry also argued that profits should be related to turnover rather than capital and that fixed prices were preferable to target prices because they provided contractors with a greater incentive to improve efficiency.[111] The SBAC was, however, by no means completely inflexible. On the contrary, 'most members were of the opinion that there could be a reduction in the customary rate of negotiated profit, provided that there were a more adequate protection in regard to the future.' In return for a lower rate of profit the industry hoped to secure improved depreciation allowances.[112]

No such concessions were forthcoming. There were, however, senior costing staff at the Air Ministry who agreed with the contractors over the advantages of the fixed price, so the Ministry now agreed to retain fixed prices and to share any losses on target price contracts. A new profit formula cut the ratio of profit to capital by reducing the rate of profit on turnover as turnover increased. The greater the difference between the industry's receipts in 1936–38 and its receipts after 1 March 1939, the lower was the rate of profit allowed from then on.[113]

By this time the political objections to statutory economic controls had been undermined by the worsening international situation and the necessary legislation, the Ministry of Supply Bill, was shortly to be presented to Parliament. The SBAC consequently came under intense pressure from the Air Ministry to accept the terms on offer. To the industrialists, however, it seemed that the government had given very little away. The new profit formula was objectionable because it implied that production had been restricted between 1936 and 1938 by a shortage of industrial capacity, and that output then expanded because of state capital investment. This was not the case. In 1938, the Air Ministry had itself concluded that ample floor space and machining capacity existed in the industry but that 'gaps in production were really caused by lack of design capacity'.[114] In other words, the constraints on the industry's turnover in the early rearmament years were largely of a technical nature and a greater output would not have been achieved simply by investing more capital.

The industry was particularly opposed to the application of the new formula to payments received in 1939, the year in which firms like Vickers and Supermarine were expecting to recover many of the production costs incurred in 1937 and 1938. The manager of Vickers-Armstrong's aviation division, Alexander Dunbar, felt that 'where the industry, out of its own resources, had extended its buildings, plant and machinery, they were entitled to a normal rate of profit . . . Thereafter there could be a reduction.' It was nevertheless decided that the new sliding scale of profit offered some scope for a settlement if somewhat higher rates of profit could be agreed.[115]

McLintock therefore proposed a compromise sliding scale which was accepted by Kingsley Wood on 24 May. In return the industry then sought longer-term safeguards which the Air Ministry went some way towards granting.[116] By the middle of June the second agreement had been finalised. The most significant gain for the industry concerned the commencement of the new arrangements which were now to apply to *contracts placed* after March 1939 rather than to *payments made* after that date.[117] In other respects, however, the 1936–38 profit standard still favoured the government.[118]

At this stage the details were published of a new tax on profits, the Armaments Profits Duty (APD), which was part of the Ministry of Supply Bill. APD was to be levied on earnings in excess of a standard set by the profits of 1935–37. The greater the difference between the profits earned in these years and the profits earned after 1939, the greater would be the amount of profit subject to taxation. In other words, APD operated in a manner very similar to the profit formula contained in the new McLintock Agreement.[119]

The industry was not expecting the new agreement to be applied in addition to the APD. When it became clear that they would be subject to the duty, the contractors felt cheated and agreed that further concessions should be sought.[120] The Air Ministry retorted, however, that the industry would make substantial gains from the concessions already made and that nothing further could be contemplated. On 7 July McLintock therefore accepted the agreement on behalf of the SBAC.[121]

THE OPERATION OF THE SECOND MCLINTOCK AGREEMENT

The original McLintock Agreement, as revised in 1938, failed to serve the government's political interests because it enabled the aircraft industry to earn excessive profits. The second agreement still allowed firms to realise relatively high pre-tax profits on their own capital.[122] Yet by combining the new agreement and the APD the government came dangerously close to placing the political need to limit profits before the military necessity of increasing production. While the state itself stood to gain from the operation of factories on commercial lines (because of the incentive for increased efficiency) some companies now sought alternative forms of remuneration. 'It is apparent', Dunbar wrote, 'that considerable difficulty would be experienced in arriving at either a fixed price or a target price which would be mutually satisfactory.' He therefore suggested that the new Wellington factory at Chester should be managed on an agency basis.[123] Similar requests were made by other contractors following the introduction of the Excess Profits Tax (EPT – a tax on all businesses which replaced APD) in 1940.[124] The difficulty here was that, by eliminating the profit motive, agency agreements also dispensed with the only practical inducement for contractors to cut their costs.[125] In order to make commercial operation more attractive the EPT was adjusted in industry's favour in 1941.[126]

The decision to retain fixed prices and to moderate the impact of the EPT finally allowed a workable pricing mechanism to emerge. The government remained concerned about the level of pre-tax profits earned by the aircraft industry, but insofar as high profits on capital resulted from very low profits on turnover it was difficult to impose further restrictions. In 1940–41, for example, 19 leading aircraft firms received government profit allowances averaging just 4.4 per cent on turnover, which allowed them to earn an average of 19.3 per cent on their capital.[127] Bombing, factory dispersal, labour dilution and frequent design changes ensured that no contractor in the industry would accept a lower rate on turnover. As one MAP official remarked:

The aircraft firms are still anxious . . . to secure fixed price settlements; these give them a chance of reward for efficient production . . . Very naturally, however, they hesitate to accept such settlements unless they believe that the price contains a margin to insure them against the contingencies which affect production at the present time.[128]

In short, low profit allowances could penalise some firms for production problems which were not of their own making, but the general application of higher rates could also allow others to realise excessive profits. This is again illustrated by the figures for 1940–41. In this period the actual profits of the leading aircraft contractors exceeded the government's estimates by an average of 1.9 per cent on turnover but in five cases actual profits were lower than the government estimates, while in four others they were considerably higher.[129]

Such inconsistencies resulted in the termination of the second McLintock Agreement in June 1941, but attempts to secure a new contractual settlement proved fruitless and were not pursued with great conviction. Instead, the government's profit estimates in fixed prices were progressively reduced from an average rate on turnover of 4.4 per cent in 1941 to 2.9 per cent in 1944. In response, the majority of firms undercut the government's estimates, so earning an extra 2–3 per cent on turnover or 5–6 per cent on capital.[130] So if, once again, the government failed to prevent its contractors from realising high profits, it did at least ensure that profits were linked to gains in efficiency which ultimately benefited the war effort.

There were two strong upward pressures on production costs during the war. First, as operational requirements became more and more exacting, aircraft became heavier and more complex.[131] Second, there was a steep rise in engineering wages, average weekly earnings increasing by nearly 100 per cent between October 1938 and July 1944. Yet, according to one contemporary estimate, the average price per pound (weight) of aircraft produced in Britain fell by about a third between 1938 and 1941.[132] The labour and sub-contract cost of the De Havilland Mosquito fell from £6,000 in 1941 to £4,200 in 1945. The first production Lancaster at A.V. Roe cost £22,000, but by 1944 the price had fallen to £15,500.[133] More sub-stantial reductions were recorded where day work rather than piece-work was employed. The cost of the Ford Merlin engine fell from £5,640 in June 1941 to £2,484 in March 1942, and to £1,180 by the end of the war.[134] The long-term effectiveness of cost control in the aircraft industry is illustrated in Table 36.

TABLE 36

Typical Indices of Aircraft Prices, 1938–44

Type	Cumulative Delivery	Price Index	Cumulative Delivery	Price Index
Fighter 'A'	150	118	1,400	100
Fighter 'B'	4,000	104	9,750	100
Bomber	150	109	2,300	100

Source: E. Mensforth, 'Airframe Production', *Proceedings of the Institution of Mechanical Engineers*, 156 (1947), p. 38.

CONCLUSION

The step-by-step acceleration of rearmament between 1935 and 1939 gradually altered the financial relationship between the government and the aircraft industry. For much of the rearmament period the state was justified in expecting industry to play an important part in financing its operations and private investment was chiefly responsible for funding both the enlargement of capacity and the expansion of turnover between 1935 and 1939.

As the prospect of war became increasingly real, however, the government had to accept a more prominent role in financing aircraft production. New capacity planned from the summer of 1938 onwards was largely financed by the state and managed on an agency basis, but as the volume of production increased so did the government's role in providing working capital to firms employed on commercial contracts. Substantial sums were still financed by the private sector in the form of bank loans and trade credit but at the same time the state liberalised the progress payment system, made special advances to the largest contractors, and apparently tolerated the accumulation of very considerable arrears in the payment of EPT and other taxes to the Inland Revenue.

The shift in the balance between state and private finance occurred relatively painlessly; over-capitalisation would have benefited neither the government nor the aircraft industry in the long term. In contrast, the search for a general contractual agreement proved much more contentious. Such an agreement had necessarily to serve the interests of both contracting parties. The McLintock Agreement did so to the extent that it encouraged the industry to expand capacity and increase turnover so that it could meet the demands of the defence programme. In other respects, however, the agreement was much less successful. Facing acute technical problems and a slowdown in turnover, the industry insisted on a time-and-line provision which was initially accepted by the Air Ministry and then rejected by the government as a whole. The revision of the agreement in the spring of 1938 seemed, for a time, to satisfy both state and industry, but the expansion of output ultimately enabled the firms to earn profits which the government considered excessive, and the McLintock Agreement was terminated as a result.

Nevertheless, although an acceptable contractual settlement proved elusive between 1936 and 1939, a relatively effective pricing formula was developed in the following two years. This involved compromises on both sides. The industry was compelled to accept lower profit allowances and a profits tax. The government continued to pay fixed prices (which it had tried to abolish in 1939) and in 1941 reduced the severity of the EPT. If contractors managed to undercut fixed prices, they could expect to retain a reasonable proportion of their earnings but the resulting improvements in productivity would benefit the war effort as a whole.

NOTES

1. Hornby, *Factories*, p. 212.
2. PRO AVIA 46/72, statement of annual expenditure on aircraft and associated products, 1931–44.
3. Issued capital of these three companies amounted to £1.5 million. See *The Statist*, 29 June 1935, 20 July 1935; Vickers 61, statement submitted to the Royal Commission on Private Manufacture of and Trading in Arms, 1935. The nominal capital of Vickers Aviation was increased from £200,000 to £900,000 in November 1935. See Vickers K757, unsigned minute, 19 November 1935.
4. *The Statist*, 6 April 1935.
5. Weir 19/1, Weir to DTD, 21 July 1937. Lazard's approached Weir in the hope of obtaining more work for Napier.
6. Data compiled from firms' published accounts in *The Statist*, 1935–39; Fearon, 'Handley Page', p. 81.
7. *The Statist*, 29 June 1935.
8. *The Statist*, 17 July 1937. This sum was credited to the company's reserve in addition to £200,000 allocated from profits.
9. *The Statist*, 20 July 1935, 22 February 1936, 6 June 1936.
10. Data compiled from firms' published accounts in *The Statist*, 1935–39.
11. HSA, Blackburn board minutes, 3, 4 and 5 June 1935, 18 July 1935, 2 April 1936.
12. PRO AIR 6/23, S of S EPM, 15 October 1935; *The Statist*, 27 July 1935.
13. PRO AIR 6/23, S of S EPM, 29 October 1935, 19 and 26 November 1935.
14. *The Statist*, 30 March 1935.
15. *The Statist*, 25 May 1935, 1 June 1935.
16. *The Statist*, 2 May 1936.
17. Weir 19/23A, Air Ministry statement on employment in the aircraft industry, 30 June 1936.
18. *The Statist*, 17 July 1937.
19. None is listed in the Ministry of Aircraft Production monthly statistical bulletin, although British Marine subsequently reappeared as Folland Aircraft.
20. Lloyd, *Years of Endeavour*, p. 159; HSA, De Havilland board minutes, 27 May 1940.
21. PRO AVIA 10/154, 1st DUS to Nelson (managing director of English Electric), 6 August 1938.
22. PRO AVIA 10/169, memorandum by AMDP, 11 March 1940.
23. Postan, Hay and Scott, *Design and Development*, pp. 72, 97; Sir Eric Mensforth, *Family Engineers* (London, 1981), p. 73.
24. Mensforth, *Engineers*, p. 73.
25. AC 70/10/55, Acland (Westland) to Handley Page, 25 May 1937; Mensforth, *Engineers*, p. 73.
26. N. Shute, *Slide Rule: The Autobiography of an Engineer* (London, 1954), pp. 166–7, 186, 192, 196, 198.
27. Ibid., pp. 199–201.
28. Ibid., pp. 235–6, 238, 247; 8751 Oxfords were eventually built, 4961 by Airspeed.
29. HSA, Airspeed policy committee minute book, meetings of 26 January 1940 and 28 March 1940; De Havilland board minutes, 27 May 1940.
30. D. Thoms, *War, Industry and Society: The Midlands, 1939–45* (London, 1989), pp. 15–17, 80–5.
31. Figures compiled from *The Statist*, 1936–38.
32. *The Statist*, 20 November 1937.
33. The actual proceeds of aircraft share issues would have been considerably greater than this. Unfortunately share premiums were not always shown separately in the published accounts.
34. Ashworth, *Contracts*, p. 199.
35. *The Statist*, 28 December 1935.
36. Bullock papers, the Secretary to US of S, 31 May 1933.
37. *The Statist*, 26 November 1938.
38. AC 70/10/79, Handley Page to A.V. Alexander MP, February 1939.

39. PRO AIR 6/43, meeting between the Air Ministry's Advisory Committee on Contracts and the SBAC, 24 October 1935; Postan, *War Production*, p. 450.
40. PRO AIR 2/2572, D of C to PAS and the Secretary, 2 February 1937.
41. HP 8, chairman's speech to the Handley Page annual general meeting, 21 July 1937; HP 759, notes for the Handley Page annual general meeting, 26 July 1939; Handley Page accounts, 1937 and 1939.
42. Vickers 1371, Vickers Ltd minute book of board meetings, 1935–37; Vickers 324, statements of order books, 1936–38.
43. *The Statist*, 2 January 1937, 18 December, 1937.
44. HSA, Blackburn board minutes, 18 June 1937, 9 September 1937, 3 December 1937, 23 June 1938, 28 October 1938.
45. PRO AIR 2/2572, D of C to First DUS, 22 July 1939.
46. AC 70/10/55, Acland to Handley Page, 25 May 1937.
47. PRO AIR 6/43, meeting between the Air Ministry's Advisory Committee on Contracts and the SBAC, 24 October 1935; PRO AIR 2/2572, D of C to PAS and the Secretary, 2 February 1937; D of C to 2nd DUS, 14 April 1938.
48. PRO AIR 2/2572, DDC (2) to PDDC, 17 October 1938; DDC (2) to D of C (A), 27 July 1939.
49. PRO AIR 2/2572, Handley Page to DDC, 6 March 1940; D of C (2) to F6, 11 March 1940; ADCF to DCF, 14 June 1940; PRO AVIA 10/170, Handley Page to US of S, 21 March 1940; AC 70/10/169, company secretary to managing director, 3 May 1940.
50. Vickers 626, Craven to Dunbar, 3 November 1939.
51. PRO AIR 20/2379, Sir Harold Howitt to First DUS, 2 February 1940; PRO AVIA 15/3710, Daniel (Bristol Aeroplane Company) to DDC, 13 November 1940.
52. Ibid.; PRO AVIA 15/3710, D of C to FUS, 2 December 1940.
53. HSA, De Havilland board minutes, 18 November 1941, 4 August 1942.
54. PRO AIR 2/2572, E.G. Compton (Treasury) to D of C, 1 April 1940; Compton to DDC, 18 April 1940; DDC (2) to D of C (A), 27 July 1939; DDC (2) to F6, 11 March 1940; statement prepared by DCF, D of C and D of C (A) (undated but approximately June 1940); PRO AVIA 15/3710, D of C to FUS, 2 December 1940. By April 1942, £5 million had been advanced to the Hawker Siddeley group. See PRO AVIA 15/3718, note by PS9, 3 April 1942.
55. Postan, *War Production*, pp. 451–2.
56. HSA, De Havilland board minutes, 2 July 1940.
57. Ashworth, *Contracts*, pp. 199–201.
58. Ibid., pp. 192–3.
59. R.S. Sayers, *Financial Policy, 1939–45* (London, 1956), p. 211. Ashworth, *Contracts*, p. 200, gives a figure of £2,015,000 for 'undistributed profits and reserves employed in the business' for the rearmament years. This is certainly too low. In the published accounts, on which his figures were based, no distinction was made between trade creditors and taxation reserves, although the latter were, in effect, undistributed profits.
60. PRO AVIA 10/245, memorandum by MAP (signature illegible), 31 January 1945.
61. Ashworth, *Contracts*, p. 196. It was, of course, for the Inland Revenue rather than the MAP to curb these practices. Unfortunately the relevant Inland Revenue file on aircraft industry profits and reserves is closed until 2020.
62. PRO AVIA 15/1040, statement by the MAP of profits in relation to capital employed, 1940–41 financial year.
63. PRO AVIA 15/3718, unsigned letter from the MAP to the Treasury, 8 July 1941.
64. Shay, *British Rearmament*, pp. 117–18.
65. Weir 19/16, note by Weir, 27 May 1935.
66. Assurances were given by the Prime Minister in May 1935. See Shay, *British Rearmament*, 103.
67. AC 70/10/159, managing director to works manager, design engineer and production engineer, 1 August, 1935; Vickers 1392, note by McLean, 11 February 1935; *The Statist*, 6 April 1935.
68. Ashworth, *Contracts*, pp. 201–2.
69. Technical costing entailed the estimation of production costs by qualified engineers and accountants employed by the government departments.
70. Weir 19/16, memorandum by D of C, 4 June 1935.

71. AC 70/10/54, meeting of the SBAC council, 8 October 1935; note by McLean, 30 March 1936.
72. AC 70/10/54, Handley Page to Fairey, 15 November 1935.
73. AC 70/10/54, Allen to McLean, 29 November 1935; Handley Page to McLean, 31 March 1936; note by McLean of a meeting with Sir Hardman Lever, chairman of the Air Ministry's Advisory Committee on Contracts, 30 March 1936; meeting between the Air Ministry and the SBAC, 19 March 1936. The number of technical costs officers and accountants increased from 11 in 1934 to 136 in 1938. See PRO AVIA 10/241, memorandum by D of C, 17 February 1938.
74. AC 70/10/54, note by McLean of a meeting with Sir Hardman Lever, 30 March 1936; McLean to Handley Page, 30 March 1936; Allen to McLean, 2 April 1936.
75. PRO AIR 6/45, meeting between the Air Ministry and McLintock, 20 May 1936.
76. AC 70/10/54, report by McLintock, 27 May 1936.
77. PRO T 161/1323/S.40700/42, statement by the Air Ministry of Capital Clause commitments, 28 February 1938; HSA, Blackburn board minutes, 18 May 1936, 9 September 1937.
78. AC 70/10/54, Handley Page to McLean, 31 March 1936; Handley Page to McLintock, 28 March 1939.
79. See Chapter 3.
80. PRO AIR 2/9165, McLintock to S of S, 4 August 1936.
81. Shay, *British Rearmament*, pp. 121–2.
82. AC 70/10/54, note by McLean of a meeting with Sir Hardman Lever, 30 June 1936.
83. PRO AIR 2/9165, McLintock to S of S, 28 September 1936; report by McLintock to the SBAC, 28 September 1936; S of S to McLintock, 7 October 1936 (three separate letters).
84. PRO AIR 2/9165, McLintock to S of S, 8 October 1936.
85. PRO AVIA 10/241, memorandum by Second DS, undated but approximately October 1937.
86. PRO AVIA 10/241, meeting between Second DS, Contracts, Accounting and Technical Cost staff, 1 November 1937.
87. AC 70/10/54, Handley Page to McLintock, 28 March 1939.
88. Calculated from Air Ministry and Ministry of Aircraft Production price books; AC 70/10/139, statement of Hampden costs, Handley Page to company secretary, 20 December 1940.
89. Vickers 339, Boddis (Supermarine) to Yapp (Vickers-Armstrong), 7 November 1938.
90. Vickers 324, Reid-Young (Vickers) to the Secretary of the Air Ministry, 18 May 1938.
91. PRO AVIA 10/242, the Secretary to second DS, 15 January 1938; Vickers 324, Reid-Young (Vickers) to the Secretary of the Air Ministry, 18 May 1938.
92. Lloyd, *Years of Endeavour*, Table XI, p. 231.
93. HSA, Hawker accounts, 31 March 1936 and 31 March 1938. Detailed accounts for 1937 were unavailable but net profits in that year were lower than in 1938.
94. Vickers 315, Identified Board Documents, 1936–38; Vickers 1798, report and accounts to 31 December 1936; Vickers 1799, report and accounts to 31 December 1937; Vickers 1371, Vickers Ltd Minute Book of Board Meetings, 1935–37.
95. PRO AIR 2/9167, note on aircraft contracts by Deloitte, Plender, Griffiths and Co., Chartered Accountants, 23 January 1939.
96. PRO T 161/922/S.40730/04, Gregg (Inland Revenue) to Gilbert (Treasury), 1 December 1938.
97. PRO AVIA 10/241, statement of airframe prices by DDC (1) 12 April 1938.
98. Ibid; Ashworth, *Contracts*, p. 118.
99. AC 70/10/54, Allen to Handley Page, 12 October 1937.
100. AC 70/10/54, Handley Page to the SBAC, 18 November 1937; PRO AIR 2/9165, note by Second DS, 31 January 1938. The 'basic price' offered less profit than the 'maximum price' but effectively dispensed with all risk of loss to the contractor.
101. AC 70/10/54, Allen to the SBAC negotiating committee, 14 January 1938; meeting between the Air Ministry and the SBAC, 17 January 1938; Allen to Handley Page, 3 February 1938; Allen to Handley Page, 5 April 1938.
102. AC 70/10/54, Handley Page to McLintock, 28 March 1939.
103. Weir 19/1, Disney to Weir, 2 July 1937.

104. Shay, *British Rearmament*, p. 116.
105. PRO T 161/922/S.40730/04, Gregg to Gilbert, 1 December 1938; HP 759, chairman's speech to Handley Page's Annual General Meeting, 21 July 1937; Shay, *British Rearmament*, pp. 255–6.
106. Shay, *British Rearmament*, p. 256; Ashworth, *Contracts*, p. 87; Hornby, *Factories*, p. 196.
107. Shay, *British Rearmament*, pp. 256–7; PRO AIR 2/9168, meeting between the Air Ministry and the SBAC, 1 March 1939.
108. PRO T 161/922/S.40730/04, Gregg to Gilbert, 1 December 1938; PRO AVIA 10/253, interim report on aircraft contracts by the Treasury Committee on Contract Procedure, 10 March 1939. Ironically the Treasury Committee now recommended that 'the costed contract should be the rule'.
109. AC 70/10/54, memorandum by the Air Ministry, 1 March 1939; PRO AIR 2/9168, meeting between the Air Ministry and the SBAC, 1 March 1939.
110. AC 70/10/54, memorandum initialled EAE/BMH, 1 March 1939.
111. AC 70/10/54, Handley Page to McLintock, 28 March 1939; meeting between McLintock and the SBAC, 29 March 1939. Fixed prices gave all savings to the contractor. Savings on target prices were shared between the contractor and the Air Ministry.
112. Ibid.
113. AC 70/10/54, PUS to McLintock, 18 April 1939.
114. PRO AIR 6/54, memorandum by DGP, 5 September 1938; PRO AIR 6/35, S of S EPM, 19 July 1938, 14 September 1938.
115. AC 70/10/54, meeting of the SBAC, 4 May 1939.
116. AC 70/10/54, PUS to McLintock, 12 May 1939; meeting between McLintock and the SBAC, 24 May 1939; PUS to McLintock, 6 June 1939.
117. AC 70/10/54, S of S to Bruce-Gardner, Bruce-Gardner to Handley Page, 12 June 1936; note of a meeting between the SBAC and the Air Ministry, 14 June 1939.
118. The Ministry used firms' receipts rather than turnover to set profit standards for 1936–38. Costs incurred by firms in 1938 but not recovered until 1939 were thus excluded from the profit standards, although a higher rate of profit was now allowed on such costs
119. Ashworth, *Contracts*, p. 84.
120. AC 70/10/54, meeting of the SBAC, 21 June 1939; PRO AIR 2/9168, meeting between the SBAC and the Air Ministry, 21 June 1939.
121. AC 70/10/54, meetings between the SBAC and the Air Ministry, 6 and 7 July 1939.
122. Shay, *British Rearmament*, p. 262.
123. Vickers 389, Dunbar to second DUS, 17 March 1939.
124. Ashworth, *Contracts*, p. 85; PRO AVIA 10/253, meeting between Sir Frank Spriggs and the Deputy Secretary, MAP, 18 March 1941. Spriggs was persuaded to continue operating Hawker Siddeley factories on a commercial basis.
125. PRO AVIA 10/247, Sir Archibald Rowlands (Permanent Secretary, MAP) to Sir Richard Hopkins (Permanent Secretary, Treasury), 16 November 1940.
126. Ashworth, *Contracts*, p. 84.
127. PRO AVIA 15/1040, statement by MAP of aircraft industry profits in relation to capital employed in 1941.
128. PRO AVIA 15/1040, Assistant Secretary to Permanent Secretary, MAP, 30 December 1940.
129. PRO AVIA 15/1040, statement by MAP of aircraft industry profits in relation to capital employed in 1941.
130. *Public Accounts Committee, Second Report* (London, 1944), Appendix 15.
131. For example, the Spitfire's weight increased from 2000 lb in 1939 to 2300 lb in 1943–44.
132. Mensforth, 'Airframe Production', pp. 37–8. Between 1938 and July 1944 average weekly earnings for male workers increased from 83*s.* to 155*s.* 6*d.* For female workers the equivalent figures were 40*s.* and 79*s.* 2*d.*
133. MAP price books. De Havilland Mosquito costs are the average for 1941.
134. G. Mills, 'Ford and the Merlin (part two)', *The Archive*, Vol. 9, No. 28 (1991), p. 71.

7

British aircraft production, 1939–41

Before the outbreak of war, Air Ministry planning aimed to give Britain the capacity to produce 2000 aircraft per month by the end of 1941. On 22 September 1939, however, the government approved a new monthly goal of 2550 aircraft for the third year of hostilities; this came to be known as the 'Harrogate' programme.[1] This ambitious target was based on the assumption that under wartime conditions it would be possible to accelerate output much more rapidly than in peacetime. There was, however, much about wartime production that was very difficult to foresee. Was there a sufficient balance between the different sectors of the aircraft industry to sustain the projected rate of expansion? How much faster could labour be absorbed? How soon would new capacity become productive? Moreover, the industry was on the point of entering a new phase of technical instability. In September 1939 it was impossible to predict how successful the future design programme would be or to quantify the loss of production which would occur while new types were being introduced.

In the event, many of the pre-war assumptions were revised, and between September 1939 and December 1941 Britain's aircraft production plans were radically altered. The quality of aircraft was deliberately sacrificed for quantity during the Battle of Britain. Practical experience proved the value of some aircraft and exposed the shortcomings of others. Factories were bombed and dispersed at the end of 1940. The test of wartime acceleration revealed imbalances and inefficiency within the industry which had to be corrected. Strategic aims were clarified; factory capacity and production programmes had to be adjusted accordingly. This chapter examines how the pre-war plans were adapted during the first two years of hostilities into a form better suited to long-term wartime requirements.

MOBILISATION PROBLEMS

The deliberations surrounding the Harrogate programme in September 1939 exposed deep divisions within the aircraft industry. Some industrialists saw a definite role for the SBAC in co-ordinating the programme on the basis of available statistics about industrial capabilities, and believed that negotiations between the Air Ministry and individual firms would simply lead to competitive promises which would prove impossible to realise. Larger companies like Bristol and Hawker Siddeley preferred individual negotiations with the Air Ministry.[2] So when they met Freeman on 18 October to consider the new programme, the firms were unable to offer a clear agenda for discussion.[3] It was argued by some that it was impossible to fulfil the plans and that they would provide no scope for flexibility if wartime requirements changed. 'It seems this programme cannot be carried out within the required time,' Fairey wrote, 'and if the true facts were now known of what the output would be in say two years time it is more than likely a different strategy would be adopted.'[4]

Yet the society could not agree on any proposals for improving the programme as a whole.[5] As Handley Page commented: 'The industry says it cannot carry out the programme put up to it and leaves it to the Air Ministry to find a solution – a most deplorable situation for an industry which claims it knows its job and can advise on the best way to get production.'[6] A few days later he reiterated this point to Bruce-Gardner: 'The situation in regard to the future is serious and calls for the industry to help AMDP [Freeman] and DGP [Lemon] in the solution of their difficult problem.'[7] Yet Bruce-Gardner supported Bristol and Hawker Siddeley, to the dismay of Richard Fairey:

Again our well-meant efforts to stop this sort of thing and to give instead constructive proposals as to how [a solution] could be achieved look like being defeated by the fact that whereas some of us take a certain view of it, Bristols would like to be left in their parochial seclusion, and Spriggs still rather has his order book in his mind.[8]

So negotiations on the Harrogate programme proceeded with companies on an individual basis. The Air Ministry clearly felt justified in setting very high production targets, for output was expanding rapidly in 1939. Events, however, were soon to support the scepticism of industrialists like Handley Page and Fairey.

The expansion of the airframe sector began to impose immense pressure on the supply of materials and equipment. In September 1938 Lemon had calculated that light alloy supplies would be sufficient for the programme[9] but he assumed that available material would be evenly distributed between the contractors whereas, until the outbreak of war, aluminium was sold in a piecemeal fashion which favoured the earliest bidders. By July 1939 Lemon was warning that 'the widening of the airframe production front and the introduction of several hundreds of subcontracting firms have tied up part of the available supplies in transit, store and process, with adverse effect upon the nett amount in course of assembly'.[10]

Thus, on the eve of hostilities, with much of the war potential capacity for materials and equipment far from complete, serious deficiencies began to arise. In July 1939 Lemon referred to 'shortages of embodiment loan equipment [which] were having an adverse effect on aircraft deliveries'.[11] In September, Spitfire production was disrupted by a shortage of engine accessories and Freeman had to intervene personally to divert the necessary supplies to Supermarine.[12] By January 1940 light alloy extrusions had become the limiting factor and, in February, the responsible aluminium controller warned that in the following months 'the supply of extrusions would barely satisfy demands'.[13]

The Air Ministry had assumed that production would accelerate following the outbreak of war. Labour intake and working hours were both expected to increase, factories would operate a two-shift system and subcontracting would be extended.[14] In reality the extrusion problem imposed such severe restrictions that, between January and March 1940, the labour intake nearly came to a standstill.[15] In January Lemon remarked that the existing work-force was not working at full pressure because of material shortages, and some machine tools were also idle. By February these deficiencies 'were unfortunately causing a temporary cessation of employment for certain operatives'.[16]

Aircraft deliveries did not live up to expectations for several other reasons. The call-up of reservists disrupted output. Vickers, for example, lost 8 per cent of their personnel to the armed services.[17] Several factories were changing over to new types of aircraft which inevitably involved some temporary loss of production.[18] There were fewer working days in December because of the Christmas break. Adverse weather conditions held up flight trials during February and an outbreak of influenza caused high absenteeism throughout the industry.[19] But the most important factor of all was that the RAF began to find itself short of spare parts, so the aircraft industry was directed to increase production of spares, if necessary by sacrificing the output of complete aircraft.[20] The result is illustrated by a comparison of actual aircraft deliveries with the Air Ministry's predictions for a war commencing in October 1939, and with the Harrogate programme (see Table 37).

TABLE 37

Comparison of Aircraft Production Programmes and Deliveries, October 1939 to April 1940

	War Potential Programme	Harrogate Programme	Actual Deliveries
October 1939			748
November			795
December	1,178		600
January 1940		977	802
February		1,001	719
March	1,413	1,137	860
April		1,256	1,081

Source: M. Postan, *British War Production*, pp. 473–4, 484.

Although output increased significantly in comparison with the first half of 1939, the Ministry was unable to meet its initial wartime targets, and as a result it found itself facing mounting political criticism.[21] The Select Committee on National Expenditure interested itself in individual projects like the Castle Bromwich Spitfire factory.[22] The firms themselves became restive as efforts to improve deliveries were frustrated by circumstances beyond their control,[23] and successive Cabinet discussions of Germany's superiority in first-line air strength confirmed the opinions of the First Lord of the Admiralty and future Prime Minister, Winston Churchill, that the Air Ministry's production branch was characterised by 'muddle and scandal'.[24]

At the beginning of April 1940 Lemon was replaced by the managing director of Vickers-Armstrong, Sir Charles Craven, who became Civil Member of the Air Council for Development and Production as Freeman's equal. This new arrangement had only existed for a month, however, when Chamberlain's administration collapsed and Churchill's coalition came into being. The aircraft industry's output had increased by nearly 50 per cent between March and May 1940 but this could not dispel the new premier's conviction that the existing Air Ministry organisation was incompetent, that aircraft were so complex that a special department was required to administer their production, and that such a department could not remain under the control of the Air Ministry without overwhelming the latter's military functions.[25] On 17 May, therefore, the Air Ministry relinquished its jurisdiction over the Department of Development and Production, and the Ministry of Aircraft Production was formed under Lord Beaverbrook.

THE AIRCRAFT PROGRAMME TRANSFORMED

On the outbreak of war, extensive group production arrangements were being made for three heavy bombers – the Stirling, the Halifax and the Manchester – and capacity for two established medium bombers, the Wellington and the Hampden, was being enlarged. In the fighter sphere, the RAF would depend on the Spitfire and the Hurricane for the immediate future, but planning for 1941 was chiefly focused on Hawker's Typhoon. It was anticipated that the Short Sunderland flying boat would soon be replaced by the Saunders-Roe Lerwick (although no major industrial commitment to the latter aircraft had been made); the Blackburn Botha was to fulfil Coastal Command's requirements for a torpedo-bomber, and the Fleet Air Arm had selected the Fairey Albacore as a successor to the Swordfish torpedo-bomber and the Blackburn Skua as a naval fighter.

The production plans were substantially revised during the next two years as the official development programme was dogged by technical problems. The development timetables for the Hawker Typhoon and for its Sabre engine had been underestimated, and when the aircraft entered service it proved inferior to the latest Spitfires. The Sabre's shortcomings caused such insuperable production difficulties that the resources of the new Liverpool factory were never very efficiently employed, but the Typhoon was eventually produced at Gloster throughout the war after a ground attack version of the aircraft proved highly effective.[26]

The Typhoon represented a small industrial commitment by comparison with aircraft like the Spitfire. The industrial cost of the failure of the Stirling bomber was far greater. The Air Ministry had always taken the view that at least one of the new four-engined types was likely to fail and that resources would then have to be concentrated on the more successful designs.[27] In the meantime, however, three factories were allocated to the Stirling, and when its technical deficiencies became clear these were only transferred to other types with great difficulty.[28]

The worst failures of all, however, were coastal and naval aircraft like the Lerwick and the Skua. The Lerwick's weaknesses were only disclosed after Short Brothers had dismantled the Sunderland's jigs and, as a result, production of Britain's only long-range flying boat was severely restricted for much of 1940. Similarly the Fairey Albacore proved obsolete on its entry into service and plans to replace the Swordfish had to be shelved.[29] Indeed, manufacturing arrangements were made for a number of coastal and naval designs which proved to be of limited operational value. Production runs tended to be short and uneconomic and long delays subsequently occurred while resources were transferred elsewhere.[30]

These failures have often been blamed entirely on the aircraft firms.[31] No doubt the potential for failure was increased when major contracts were allocated to companies with limited design and production experience, like Short and Napier, but several other factors were involved. First, it was not generally appreciated that piston-engined aircraft were nearing the end of their economical development. Each new design to appear in the 1930s had offered a significant improvement in performance over existing aircraft and it was mistakenly expected that these advances would continue. Second, as already noted, the task of developing successful models was complicated by the requirements of official specifications. For example, the Air Ministry had originally demanded that the Stirling should be capable of being launched by catapult. The necessary strengthening of the airframe added considerably to its weight, thereby reducing its performance.[32] Moreover, limitations had been imposed on the aircraft's wingspan by the size of RAF aircraft hangars in the mid-1930s.[33]

In the naval sphere official requirements were particularly exacting, and the Admiralty's technical capabilities left much to be desired. As late as 1937 the Admiralty selected a biplane rather than a monoplane as a replacement for the Fairey Swordfish.

Throughout the rearmament years the Admiralty failed to recognise the potential for converting single-seater land fighters into fleet fighters. In 1939, Admiralty fighter specifications still embodied two-seater cockpits and the crucial importance of high performance in fleet fighters was only acknowledged after the commencement of hostilities. Small wonder, then, that such confused specifications as 'torpedo-spotter-reconnaissance' or 'fighter-dive-bomber' rarely produced aeroplanes of very great operational utility. Indeed, they provide the best illustrations of how a multiplicity of service requirements prepared by the military on the basis of what seemed operationally desirable (rather than technically feasible) invariably reduced the ultimate quality of aircraft design.[34]

It was, in fact, the reversal of the established procurement procedures which produced some of the best wartime aircraft; designs originating within the industry gave highest priority to technical considerations and were often based only on general ideas about operational requirements. If the basic airframe was successful it could usually be adapted for specific operational roles. The state had a responsibility for promoting design through the issue of specifications but it was by no means blind to the potential of designs proposed by the industry and it had supported many such projects in the 1930s.[35]

The success of this approach continued during the war, further confounding the expectations of the pre-war planners. For example, while Bristol's Hercules-engined Beaufighter was built to fulfil a known Air Staff requirement for a cannon fighter, it was not the product of an official specification.[36] While design and development were in progress, the RAF became increasingly interested in the aircraft's potential. Successive schemes were approved for night-fighting and Coastal Command Beaufighters, a Merlin-engined version was developed, and a group production organisation was established involving Bristol's Filton factory, one of Fairey's Stockport works, and a new assembly factory at Weston-super-Mare which relied almost entirely on subcontractors for manufacturing and sub-assembly work.[37]

The history of the De Havilland Mosquito was similar in many respects. Faced with the prospect of building another firm's design, De Havilland resurrected an idea, which had been circulating in the Air Ministry for some years, for a fast day bomber as an alternative to heavy night bombers like the Halifax and the Stirling. The Air Ministry initially expressed some scepticism and only accepted the Mosquito when early wartime experience demonstrated the need for a high-speed photographic reconnaissance aircraft. As with the Beaufighter, the true qualities of the Mosquito only became clear later on. By July 1941 orders had been placed for bomber, fighter and reconnaissance versions, and in October a group production scheme was established involving the Standard Motor Company, a Canadian factory, and a De Havilland assembly plant at Watford which was supplied with parts by the furniture industry.[38]

A.V. Roe's Lancaster bomber also owed as much to the engineering ability of the industry as to a centralised state bomber development programme. The first change from the aircraft's original P13/36 specification came in 1937 when pressure from Rolls-Royce resulted in one of the two prototypes being modified to take four Merlin engines rather than two Vultures.[39] On this occasion A.V. Roe rejected proposals that its prototype should be redesigned and Handley Page reluctantly agreed to undertake the additional work instead.[40] By the summer of 1939, however, Roe's chief designer had concluded that the twin-engined model, now known as the Manchester, was likely to fail, and a 'Manchester Mark II', equipped with four Merlins, was therefore given priority status in the company's drawing office. When, in the following July, the MAP suggested cancelling the Manchester and concentrating instead on the Halifax, A.V. Roe offered the Manchester II as an alternative. By this time most of the new design was complete. It was attractive to the MAP because it involved no new technical risks but the decisive factor was its capacity to use many of the jigs and tools which had already been ordered for the Manchester. This alone saved the government £1.7 million.[41]

Each of these aircraft was brought into service with remarkable speed despite mounting air force requirements which often complicated the task of production planning. The Beaufighter entered service just 18 months after the Air Ministry gave its support to the project. Its adaptation for Coastal Command involved 60 modifications and the manufacture of 3000 new parts but the work was completed in a matter of weeks.[42] De Havilland faced similar problems chiefly because of the Air Staff's inability to decide how many of each Mosquito variant it required. Parts of the aircraft had to be redesigned and jigs had to be modified, yet deliveries began less than two years after the inception of the design.[43] In the case of the Manchester II (renamed the Lancaster) preparations were so advanced by the time the project was approved that a prototype was built in just six months. A production order was placed in November 1940, and the first aircraft was delivered in October 1941.[44]

The fighter development programme also refused to conform to the official plans. Supermarine continued their work on the Griffon Spitfire purely on the basis of verbal instructions from Freeman,[45] although official priority had been given to the Typhoon. In the meantime, the power of the Rolls-Royce Merlin was significantly raised by a highly successful programme of supercharger development, and the prospects for improving the performance of the Spitfire were thus considerably extended.[46] So, in July 1940 Ernest Hives urged Freeman to give a greater commitment to the Spitfire III, which was powered by the new two-speed supercharged Merlin 20, and in October Hives tackled Beaverbrook. 'Anyone', he wrote, 'who suggests that Typhoons are going to make any material contribution to the Battle next year does not know what he is talking about.'[47]

There were three problems. First, the entire projected output of the Merlin 20 had been allocated to other types of aircraft like the Hurricane and the Beaufighter, which would benefit more from the increased power.[48] Second, Supermarine had not pressed forward with the development of the Spitfire III airframe, having been informed by the MAP that sufficient Merlin 20s were unavailable.[49] Third, to install the Merlin 20 into the Spitfire the airframe had to be extensively modified and this was certain to disrupt production.[50]

Nevertheless, at the end of 1940 the MAP finally realised that the Typhoon would not appear in quantity during 1941 and that an improved Spitfire would have to be introduced instead, so Rolls-Royce rapidly applied their new supercharger to a single-speed engine to produce the Merlin 45, which could be installed into the Spitfire with relative ease. The new combination, designated the Spitfire Mark V, was proposed in December 1940, and deliveries started in March 1941.[51]

The Spitfire V was less effective at low altitudes than the Spitfire III because it was powered by a single-speed engine, rather than a two-speed engine, and this was to prove a serious disadvantage when the German Fw. 190 fighter appeared in September 1941.[52] By that time, however, bombers were receiving top priority for all available Merlin 20s. It was not until July 1942 that the introduction of the Merlin 61, with its two-stage supercharger and intercooler, enabled the Spitfire to match the performance of the Fw. 190.[53] Although, in the meantime, the best German fighters proved superior to their British counterparts, Germany's advantage would have been greater still without the high degree of flexibility exhibited by firms like Rolls-Royce and Supermarine.

By 1941, then, the aircraft programme differed considerably from that envisaged by the pre-war planners. The Spitfire remained at the forefront of fighter technology; the Manchester was in the process of being succeeded by the Lancaster; three factories had been allocated to the Beaufighter, which did not feature prominently in the Air Ministry's calculations until the outbreak of war; and group production arrangements were being made for the new De Havilland Mosquito. The basis for all these developments was the Merlin engine, the potential performance of which was unrecognised in official circles until 1940.

THE BATTLE OF BRITAIN: EMERGENCY MEASURES

The Ministry of Aircraft Production

With the creation of the MAP, aircraft production gained an independent political voice which it had hitherto lacked; without it many of the emergency measures of the following months would have been impossible to implement. Moreover, the MAP's status as an autonomous department allowed the resource requirements of the aircraft programme to be more effectively co-ordinated with the government's overall industrial and strategic plans than had been the case when the Air Ministry controlled aircraft procurement.[54] Some time passed, however, before a stable administrative structure evolved within the new Ministry. Indeed, between July 1940 and 1941 the MAP was reorganised three times. The RAF's influence over production declined. The Supply Committee was renamed the Supply Board and the chair was given to Sir Charles Craven. By October only one seat was occupied by an Air Ministry representative, the Director General of Equipment, and the committee's functions had been restricted to the provision of additional factory capacity. Production questions were dealt with on an informal *ad hoc* basis by Beaverbrook's senior staff. After May 1940, airframe, aero-engine, and material production was supervised by Patrick Hennessy of Ford, while armaments, equipment and factory construction again went to Craven. The activities of RAF officers like Freeman and Tedder were confined to research and development.[55]

Freeman, Tedder and Craven left the Ministry in the winter of 1940, after which the former Vickers' manager, Trevor Westbrook, assisted Hennessy with production matters.[56] No formal body was established to supervise the MAP's affairs until Beaverbrook was replaced by Moore-Brabazon in May 1941, when an Aircraft Supply Council came into being comprising the leading production, research and development, financial, and North American supply staff.[57] Craven returned to become Controller General, with responsibility for production matters, and a controller of research and development was appointed in place of the former DGRD. Overall responsibility for production, and research and development remained divided until Freeman returned to the MAP as Chief Executive in October 1942.

During the emergency of 1940 many of the formal procedures employed by the Air Ministry's Department of Development and Production were abandoned; Beaverbrook's administration was characterised by informality, spontaneity, and improvisation.[58] Such an approach may have been necessary in the summer of 1940 but it could not have continued indefinitely without jeopardising the efficiency of the Ministry. As it was, certain aspects of MAP policy may, with hindsight, be seen as regressive. For example, while the Ministry continued to gather information about productivity and was, indeed, quite capable of acting on such information, there was no longer any formal production research of the sort commissioned by Lemon.[59] Moreover, the programming organisation created by the Air Ministry was severely weakened. Under pressure from Beaverbrook and Hennessy, the industrialists committed themselves to unrealistically high output targets; there was insufficient co-ordination between the different production directorates; decisions were taken in a vacuum, and this led to confusion not only within the Ministry but in the aircraft industry too.[60] The picture painted by Vickers-Armstrong's commercial manager was typical:

An investigation into several 'instructions to proceed' received from the Ministry discloses that as a result of amendments to programmes made from time to time, we have reached a position where we cannot be certain of the total number of "Wellingtons" of each Mark which we are expected to build.[61]

Rolls-Royce had even less faith in the official programmes. 'Everyone knows', Hives told Hennessy, 'that the . . . provisioning programmes . . . don't mean anything. We take no notice of them.'[62] Between November 1940 and February 1941, Rolls-Royce submitted four separate forecasts of Merlin deliveries for the first six months of 1941, starting at 7208 and falling to 5406.[63]

By July 1941 the weakness of the programming section had become so serious that the MAP commissioned an inquiry by John

Jewkes, a professor of economics then working for the Cabinet Office, and in November Jewkes made a number of recommendations which resulted in his appointment as head of a new Directorate of Statistics and Programmes.[64] He made his principal objective the improvement of co-ordination between the different sectors of the aircraft industry. When his directorate was established, there existed just one official programme for complete aircraft but, within a year, programmes had been drawn up for propellers, carburettors, magnetos, guns, turrets, and bombs, and additional programmes for constant speed units, propeller blades, power plants, undercarriages, and raw materials were being prepared.[65]

Most important of all, however, was Jewkes' recognition of the key role of aero-engines, and particularly the Merlin engine, in the aircraft programme. 'The engine programme should lie at the very centre of all our planning', he wrote:

Everything else must run from our views as to how many engines we are likely to produce. I take this view because the supply of engines is less flexible than the supply of other equipment; capacity for engines is more difficult to expand and the problems of finding skilled labour and critical machine tools greater than in the case of other equipment.[66]

This was a marked departure from the policy of the pre-war period. While new aircraft were still under development, aero-engines had occupied a secondary position in state planning. Jewkes now assumed, rightly, that the war would be fought with aircraft already in production, and that the RAF would rely chiefly on up-rated aero-engines, rather than new aircraft, to maintain operational efficiency. This implied that aircraft and engines would have to be continuously improved and, as one former programmer recently recalled, 'nothing was more destructive of the rhythm of production than interruption of the line to introduce some modification'. Moreover, there was a universal tendency to underestimate the loss of production involved.[67] So while the new directorate clearly made a vitally important contribution to co-ordinating the different sectors of the aircraft industry, the programmes themselves invariably proved unrealistic and successive reductions took place throughout the war. It is unlikely that the gap between production and programme could have been eliminated without qualitative sacrifices far greater than the RAF was ever prepared to accept.

Production policy

On the day Churchill's administration was formed, 12 May 1940, Freeman made several proposals for accelerating aircraft production. He advised that the introduction of the new heavy bombers should be delayed and that the firms should instead continue to produce established designs.[68] Subsequent discussions between the MAP and the Air Staff resulted in a decision to concentrate all efforts on the production of the Wellington, Whitley, Blenheim, Hurricane and Spitfire, 'at the expense of the new big bombers'.[69]

Faced with the 'Harrogate' production targets, the industry had been gradually building up a cushion of 'work in progress' with the aim of avoiding hold-ups, increasing production month by month, and then sustaining output at the required level. The process involved a time-cycle averaging six months between the production of the necessary materials and the delivery of finished aircraft. At any given time, however, this cushion, and hence the time-cycle, could be temporarily reduced and output could be increased through the application of emergency measures such as longer working hours, special transport arrangements, the requisition of spare parts from the squadrons, the reduction of spares manufacture in favour of that of complete aircraft, and the utilisation of stocks of raw materials. This was the policy now adopted by the aircraft industry with the MAP's assistance.

In December 1939, for example, the Air Ministry had instructed the industry to devote 15–20 per cent of output to spare parts,[70] but during the Battle of Britain spares production averaged less than 6 per cent of complete aircraft production.[71] The need for an immediate increase in output came to outweigh all other considerations. The aircraft industry was given absolute priority over the other munitions sectors for skilled labour. More and more sub-contractors were pressed into service.[72] New and second-hand machine tools were diverted to priority aircraft contractors from RAF depots, Army Technical Training Schools, lower priority MAP factories, and from supplies which had been destined for France.[73] The equipment of new factories was deliberately postponed so that additional machine tools could be allocated to factories building the most important five types. For example, 468 machines were transferred from the Hercules engine shadow factories to help increase the weekly output of Mercury and Pegasus engines from 150 to 200.[74]

The short-term effectiveness of such measures is illustrated by the fact that, throughout the war, the time cycle between the production of aluminium and the delivery of finished aircraft was never shorter than during the Battle of Britain. The MAP's director of materials production wrote later that 'the normal fabrication periods were telescoped to a quite remarkable degree and the material flew a very short time after it was delivered to the constructor'.[75]

The MAP was fortunate, however, in that the new capacity planned by the Air Ministry between 1938 and 1940 was coming into operation. The impact of the Air Ministry plans on light alloy production during 1940 is illustrated in Table 38.

TABLE 38

British Light Alloy Production (tons), 1939–40

Light Alloy	November 1939	October 1940
Sheet and Strip	2,200	4,850
Extrusions	1,739	2,975
Castings	2,330	2,852
Forgings	400	508

Source: PRO AVIA 10/178.

In addition to the emergency measures, the new Ministry sometimes had to intervene directly in the management of particular factories. Later, urgent requirements for heavy bomber and Fleet Air Arm types during 1942 led the government to reorganise several firms. In 1940, however, requirements were principally for fighters and while the most important factories succeeded in achieving the required increase in output there was clearly considerable scope for accelerating Spitfire production, not least from the much-delayed Castle Bromwich factory which was then under the management of the Nuffield organisation.

In November 1939, Vickers-Armstrong had been asked by the Air Ministry to increase Spitfire output from 50 to 100 per month. Arrangements were made to duplicate the production of sub-contracted parts and deliveries from the new suppliers commenced during the first months of 1940.[76] Capacity which could have been used for Spitfire production at Southampton was, however, still allocated to the Walrus seaplane in spite of the company's recommendations that Walrus production should be transferred else-where.[77] Only on Beaverbrook's instructions was seaplane manu-facture finally switched to Saunders Roe.[78] Far more capacity for Spitfire production existed at Castle Bromwich, however. On 20 May, no doubt acting on Craven's advice, Beaverbrook instructed Vickers-Armstrong to take over control of the factory from Nuffield.[79]

Vickers found that tooling at Castle Bromwich was still incom-plete and, in order to accelerate production, many machined parts had to be made by precision methods, for which there was neither plant nor personnel. The factory was also deficient in plant for the distribution of stores and the system for progressing work through the machine department had completely broken down. So great were the problems that Vickers-Armstrong eventually asked the equipment manufacturers, Joseph Lucas, to reorganise progress control in the machine department, and by July this section was operating efficiently.[80]

In the meantime, Vickers dealt with the factory's other bottle-necks. Inadequate provision had been made for wing assembly, so 24 additional jigs were ordered for this purpose.[81] It was also found that the system for recording job times was vulnerable to abuse by the work-force.[82] Castle Bromwich therefore adopted a new system which was already used in the Vickers-Armstrong group, and experienced staff were transferred from the company's other factories to help with its introduction.[83]

It may be that by the time Vickers-Armstrong assumed control of Castle Bromwich the factory had overcome many of its early problems. Nevertheless, output rose significantly following the takeover, and the factory made a crucial contribution to Spitfire production after the bombing and dispersal of Supermarine later in 1940 (see Table 39).

The supply of aircraft was further increased by the repairs organisation. Repaired aircraft accounted for nearly 40 per cent of all Spitfires and Hurricanes reaching the RAF during August and September 1940.[84] The situation here was somewhat similar to that at the Castle Bromwich factory for, while the repairs organisation had been created by the Air Ministry between 1938 and 1940, it was operating far below its true capacity at the time of the MAP's formation. Responsibility for repairs was subsequently assumed by the new Ministry, control of repairs was reorganised, and the repairs organisation was enlarged. The results are illustrated in

Table 40. If few of the emergency measures instigated by Beaverbrook's regime brought long-term benefits to British aircraft production, the expansion of the repair organisation is the most significant exception. Between 1940 and 1945, 79,000 repaired aircraft were delivered to the Metropolitan Air Force. This represented 48 per cent of total MAF deliveries.

The production figures show that the combination of the Air Ministry's plans and Beaverbrook's emergency measures helped to stimulate a remarkable acceleration during the summer of 1940. Britain produced more than twice as many single-engined fighters as Germany during the Battle of Britain and, in the year as a whole, out-produced Germany by 50 per cent.[85] Contrary to Beaverbrook's expectations, however, this rate of expansion could not be sustained indefinitely (see Tables 41 and 42). Indeed, 1941 was a particularly difficult year for British aircraft production.

TABLE 39

Spitfire Deliveries, April–December 1940

Factory	Average Monthly Spitfire Deliveries, 1940		
	2nd Quarter	3rd Quarter	4th Quarter
Southampton	77	121	59
Castle Bromwich	33	39	76

Source: PRO AVIA 10/311.

TABLE 40

Deliveries of Repaired Aircraft, 1940–41

Year	Quarter	Total Aircraft Repaired (Monthly Average)
1940	2nd	177
	3rd	660
	4th	738
1941	1st	809
	2nd	1,063
	3rd	1,336
	4th	1,312

Source: *Statistical Digest of the War*, p. 154.

TABLE 41

British Aircraft Deliveries, 1940–41

Year	Quarter	Aircraft	Structure Weight (m. lb)
1940	1st	2,381	8.9
	2nd	3,951	15.7
	3rd	4,607	18.2
	4th	4,110	16.0
1941	1st	4,515	18.7
	2nd	4,865	20.9
	3rd	5,376	23.5
	4th	5,338	24.1

Source: *Statistical Digest of the War*, p. 152.

TABLE 42

British Aero-engine Deliveries, 1940–41

Year	Quarter	Aero-engines	Thousand Horsepower
1940	1st	3,940	2,555
	2nd	6,644	4,619
	3rd	7,162	5,205
	4th	6,328	5,019
1941	1st	7,271	5,963
	2nd	8,441	6,901
	3rd	9,603	8,273
	4th	11,236	10,279

Source: *Statistical Digest of the War*, p. 155.

1940–41: THE LIMITATIONS OF EMERGENCY MEASURES

In the summer of 1939 the Air Ministry had foreseen that production of finished aircraft would fall during the later months of 1940
because the structure weight of airframes and the horse-power of
aero-engines were both scheduled to increase.[86] But several other
factors were involved. Stocks of materials used during the emergency were exhausted, fears over fatigue led to a reduction in working hours, and German bombing caused a temporary loss of output
at some factories. On 26 September, Supermarine was attacked, and

although direct damage was confined to buildings and work in progress, the continuity of production was severely upset.[87]

In the months following the raid, the number of workers at Supermarine was reduced from 3660 to 3079, many of whom were employed on salvage operations, and great difficulty was subsequently encountered in transferring skilled men to new factories.[88] The attack and its repercussions virtually halved Supermarine's output between September 1940 and March 1941.[89] Vickers-Armstrong (Weybridge) lost 'a considerable number of workpeople' because of bombing, and there was a temporary cessation of the nightshift and a reduction in overtime.[90]

The threat of bombing also undermined labour morale. In August 1940 the Castle Bromwich management estimated that the night-shift was 'not working to more than 50%–60% of dayshift efficiency' because of the workers' fear of air attack.[91]

In response, Beaverbrook adopted a general policy of factory dispersal, and by December 1941 dispersal factories accounted for 23 per cent of the airframe industry's productive floor space.[92] Dispersal rendered the industry virtually immune to air attack but it imposed its own constraints on production. The managerial consequences were noted in Chapter 5. Dispersal also increased the ratio of non-productive workers to productive workers, adding considerably to overall labour requirements. At Vickers-Armstrong's Weybridge factory, for example, the total labour force increased by 67 per cent between 1940 and 1943, but in the same period the productive labour force increased by only 35 per cent.[93]

Dispersal compounded the difficulties of integrating the subcontracting network, and subcontracting capacity had necessarily to become more self-contained, which added to machine tool requirements for 'balancing' purposes.[94] The cost of new fixed capital for dispersal was £11 million, or the equivalent of total Air Ministry capital sanctions on shadow factories between March 1936 and December 1938.[95] The overall cost in terms of both capital and production was even higher, for the policy of duplicating sources of component supply (as an insurance in case one source was bombed) imposed an enormous burden on the small tool industry and increased the tendency of individual subcontractors to accept work on several different types of aircraft. This led to inefficient use of labour and plant.[96]

Output was then reduced further by the transfer of resources to new aircraft like the Mosquito and to new variants like the

Spitfire V. 'These changes were particularly unfortunate', Supermarine's manager wrote in March 1941, 'as they occurred at the time when production was just recovering from the dispersal.'[97] The introduction of new types, deliberately postponed in May 1940, could not be delayed indefinitely without wasting capacity and jeopardising the fighting efficiency of the RAF. The most important cases, accounting for more than 50 per cent of airframe floor space in December 1941, are listed in Table 43.

TABLE 43

Factories Introducing New Types of Aircraft, 1941–42

Firm	Change From	Change To
Austin	Battle	Stirling
Short	Sunderland	Stirling
Short Harland	Hereford	Stirling
A.V. Roe	Blenheim	Lancaster
Armstrong Whitworth	Whitley	Lancaster
Handley Page	Hampden	Halifax
English Electric	Hampden	Halifax
Rootes	Blenheim	Halifax
De Havilland	Various	Mosquito
Bristol	Blenheim	Beaufighter
Gloster	Hurricane	Typhoon
Westland	Lysander	Spitfire

Source: PRO AVIA 10/311.

Finally, new factories took longer to enter production than initially expected. As had been the case between 1936 and 1938, technical problems delayed the introduction of some new types but there were also shortages of particular industrial resources such as machine tools. By 1940 the international demand for machinery was so far in excess of capacity that most factory buildings were completed before they could be equipped with the necessary plant.[98]

In June 1941, 13,000 machine tools ordered by the MAP had yet to be delivered, and another 12,000 were installed but were idle for lack of equipment.[99] An internal report on Rolls-Royce's Glasgow factory stated that 'the general machining sections require the greatest assistance as regards machining capacity and are, until more machine tools are delivered, the factors limiting any further substantial increase in engine output'. The result was a shortfall of about 15 engines per week.[100] Identical problems occurred in the

Hercules engine factories. 'Owing to the general high demand for machine tools some departure had to be made from the optimum of flow production by carrying out more than one operation on one machine.' Full 'flow production' was introduced later.[101]

During 1941, these bottlenecks reduced deliveries of equipment, components and materials to the aircraft firms and frustrated their efforts to increase production. Absolute deficiencies were, however, less of a problem than maldistribution which, in turn, resulted from the Air Ministry's decision to duplicate sources of component supply as a strategic precaution. Thus in May 1941, Fairey was short of a particular part for the Beaufighter which was in abundant supply at the two Bristol factories.[102]

EXPANSION RENEWED

So there were limits to the effectiveness of the emergency measures introduced by Beaverbrook in 1940. Nevertheless, factories sanctioned by the Air Ministry during the last year of rearmament did finally enter production during 1940 and 1941. The first deliveries were made by the new aero-engine factories, new Vickers-Armstrong and Bristol airframe assembly plants, Fairey's second Stockport works, and A.V. Roe's 'daughter', Metropolitan-Vickers. The Castle Bromwich Spitfire factory began to fulfil its true potential, and the Ford shadow factory produced its first Merlin engines at the end of 1941. So a new generation of combat aircraft entered service, including such designs as the Lancaster, Halifax, and Mosquito. The MAP's production index, which fell from 731 in August 1940 to 527 in January 1941, recovered in the following months and reached 1000 in January 1942.[103]

Throughout this period, new capital investment programmes were continuously being sanctioned. Between December 1939 and August 1941, the state capital commitment to all air orders increased from £102 million to £198 million. The bulk of this expenditure was still absorbed by airframes and aero-engines, but more and more capacity was sanctioned for airframe components and equipment.[104] New subcontracting networks were established and new factories and extensions approved for items like carburettors, magnetos, propellers, and turrets, which had all featured to a greater or lesser extent in the pre-war plans; but the state also became more actively involved in providing capacity for undercarriages, radio, radar and equipment, the demand for which had previously been underestimated.[105]

Greater provision was also made for aluminium production. The bauxite resources of British Guiana, the Gold Coast and the Dutch East Indies were tapped, production from British Guiana alone rising from 12,000 tons in 1939 to 68,000 tons in 1941. This was increasingly used to supply the Canadian virgin aluminium industry as the additional capacity sanctioned on the outbreak of war became productive. Meanwhile, projections for light alloy requirements in 1941 indicated a demand for 190,000 tons against anticipated United Kingdom production and imports of 160,000 tons, so further plans to create additional capacity were initiated, and by November 1940 no fewer than 42 different projects were in progress.[106]

Output eventually exceeded expectations, chiefly because of improvements in the technical efficiency of the producing firms and the rationalisation of production. In 1941, Beaverbrook's priority system of material allocation was abandoned and a reorganised Directorate of Materials Production introduced a more effective system based on the fabrication cycles of the aircraft required by the MAP programme. Orders were apportioned so that producers could specialise in particular types and sizes of material, and efforts continued to persuade aircraft designers to use materials which were in plentiful supply.[107] A.V. Roe replaced 66 per cent of the Lancaster's extruded alloy sections with rolled and drawn sections.[108]

The capital schemes inaugurated between 1939 and 1941 did not bring the expansion of the aircraft industry to an end, as a further £120 million was sanctioned in 1942.[109] For several reasons, a growing proportion of expenditure was allocated to machinery and plant rather than buildings. First, labour had increasingly to be substituted by capital as supplies of skilled workers were exhausted and manpower allocations cut.[110] Second, certain British aircraft were designed with a relatively large number of different detail parts in order to save weight and improve performance and machining capacity therefore invariably determined the level of output.[111] Third, deliveries of spare parts, which were relatively low in 1940, increased steadily throughout the war (see Table 44).[112] By February 1944, 18 per cent of all airframe labour was engaged in the production of spare parts.[113] The burden naturally fell on manufacturing (as opposed to assembly) capacity.

The greatest demands, however, were made by the bomber programme sanctioned by the War Cabinet in December 1941. In July,

Churchill had called for the production of 14,500 heavy and medium bombers in the following two years – an increase of 3500 over the existing target.[114] It eventually transpired that the resources considered necessary by the MAP to achieve this goal were unobtainable. Proposals to construct four new factories were therefore rejected, and instead greater reliance was placed on subcontracting for the manufacture of airframe parts and sub-assemblies.[115]

TABLE 44

Average Monthly Value (£) of Vickers-Armstrong
Spare Parts Production

Factory	July–December 1941	January–June 1944
Southampton	72,000	592,000
Castle Bromwich	29,000	354,000
Weybridge	284,000	907,000

Source: Vickers 870.

The decision to rely on subcontractors for the bomber programme has been criticised on the grounds that it imposed severe long-term constraints on the productivity of Britain's wartime aircraft economy. As the war progressed, most of the principal aircraft firms developed more advanced manufacturing processes and more rational factory organisation, but in the smaller and highly dispersed engineering workshops typical of the subcontracting network similar improvements were often much more difficult to introduce. For this reason assembly factories could sometimes process, in one shift, parts and sub-assemblies which had only been produced by the operation of multiple shifts at subcontractors.[116] Doubtless more bombers would have been produced if new self-contained factories had been sanctioned instead, but only if the necessary labour and management could have been transferred to these new factories. It is far from certain that this would, in fact, have been possible.

THE EFFICIENCY OF AIRCRAFT PRODUCTION

Although the industrial resources allocated to the aircraft industry increased steadily until the final stages of the war, continuous efforts were made to improve the efficiency with which these resources were employed. The output of Merlin gears at Rolls-Royce's Glasgow factory increased by 100 per cent following the introduction of a line system in July 1941.[117] In May 1941, a series of reforms was introduced at Bristol's Filton works following an investigation by the MAP's Technical Costs branch. A progress supervisor was appointed, with responsibility for the entire factory, to prevent the competition between different departments that had previously hindered progress control. The main stores were reorganised and improved provisions were made for storing parts close to the assembly lines. More accurate machine tool loading charts were prepared, the subcontracting and press departments were strengthened and rate-fixing was more closely monitored.[118] The Austin airframe factory was also reorganised, the wages system was reviewed, and a new layout was prepared with the aim of achieving 'the nearest possible approach to line production on aircraft'.[119]

Another investigation resulted in the appointment of a new general manager at Hawker in February 1941.[120] Until then there had been a high degree of interdependence between Hawker and Gloster, both of which built the Hurricane, but under the conditions prevailing during the Battle of Britain the output of one factory had sometimes been lowered by shortages of parts produced by the other. So when production of the Typhoon started at Gloster it was decided that the two factories should become self-contained units.[121] New machining capacity helped to reduce Hawker's vulnerability to shortages of subcontracted parts, and machining operations were replanned to improve loading economy.[122] New machining, detail manufacture, and sub-assembly facilities were also created at Supermarine and at the London Aircraft Production group (which manufactured the Halifax bomber). The MAP's Director General of Aircraft Production considered it impossible to rely entirely on subcontractors. 'There must', he declared, 'be an essential nucleus of machining capacity.'[123]

Other initiatives helped to raise the efficiency of the industry as a whole. Aluminium was used much more economically after segregation and resmelting of scrap arising from aircraft production was made compulsory in July 1940, and the relaxation of official material specifications enabled firms to use a much higher proportion of secondary material than had been tolerated in the past. In 1941, 70 per cent of aluminium scrap was segregated, and secondary material provided a third of the total amount of aluminium required for aircraft production.[124] Even greater economies were achieved by diverting resources to the repair organisation. The material used to repair 3816 Lancasters between 1942 and 1945 would have been sufficient to produce only 622 new airframes.[125]

At the manufacturing end of the production line there were further advances in the application of press work. By 1941 the value of the hydraulic rubber platen press for metal-forming had been proved at a number of aircraft firms,[126] and two British press manufacturers were co-operating in the design of a standard hydraulic press for aircraft production. A 2500-ton press based on the unit principle was soon developed; its capacity could be doubled or tripled simply by the addition of identical hydraulic units.[127] In July 1941 (by which time 14 presses were in operation at such firms as Blackburn, A.V. Roe, and Metropolitan Vickers) the MAP decided to place a bulk order for 16 of the standard presses with a total capacity of 41,000 tons.[128] Properly set up, these machines proved capable of producing over 20,000 parts per week.[129] Further bulk purchases of labour-saving equipment, like the cecostamp drop hammer, were made in the following year.[130]

Other complex operations were accelerated by the introduction of special machinery. The time taken to machine the Lancaster's wing spars was considerably reduced by the employment of a device known as the plano-miller, the efficiency of which was itself improved by more than 90 per cent during the war.[131] Special stretching and milling equipment was also provided at Castle Bromwich to manufacture the Spitfire's wings.[132] The replacement of handwork by such machinery allowed the time required to make fuselage formers for one type to be reduced from seven hours to 20 minutes.[133]

Efforts were also made to simplify the design of aircraft components which had proved difficult to construct. The installation of hydraulic pipes into the bomb compartment of A.V. Roe's Manchester absorbed about 105 man-hours, but by reorganising the pipe runs into a separate sub-assembly, the installation time for the Lancaster was cut to five hours.[134] The Manchester's fuel tank was redesigned to be spot-welded, rather than riveted, to the Lancaster airframe.[135] The application of this method to secondary structures could cut production times by as much as 50 per cent.[136] At Rolls-Royce a 'production development' team introduced several major changes to the Merlin engine to ease production.[137] During 1940, the Merlin was completely redrawn so that it could be mass-produced by the Ford Motor Company.[138]

Meanwhile the SBAC, which had functioned only as a trade organisation before the war, increased its involvement in technical matters by establishing a number of subcommittees through which both design and production experience could be shared. There were individual committees dealing with engines, propellers, armaments and materials, and with the standardisation of drawing office procedure; but the most successful work from the production standpoint was undertaken by the SBAC's Aircraft and Engine Design Standardisation Committee.

An SBAC standards handbook was first issued to the industry in April 1940, and at that time it contained just 30 agreed standards. By January 1943 more than 1000 standards had been incorporated into the handbook and many more individual data and reference sheets had been circulated throughout the industry. Firms were persuaded to accept standard designs when as many as 40 different ones had existed previously.[139] Work began on simple components but eventually standard power plants, rudder bars, control columns and electric wiring systems were introduced. If tooling costs for standard parts were particularly high, the SBAC itself purchased the necessary tools and amortised their cost over the quantity of components ordered.[140]

There was also a marked improvement in assembly methods. As early as December 1939 the trainer manufacturers, Phillips & Powis, installed a runway for the assembly of 'Master' aircraft. Lemon observed that

The aircraft will move forward, from one stage to the next, every four hours ... The new layout will result in a considerable reduction in cost ... When the Phillips and Powis scheme is in operation, I feel sure that other firms will be so impressed by the results achieved that they will quickly follow suit.[141]

The new arrangements reduced the Master's assembly time by 30 per cent and subsequently, as expected, the use of some form of conveyor became standard in the aircraft industry. Owing to the low rate of movement, however, it was found that simple floor trolleys provided all the economies of the mechanical conveyor at considerably less capital cost.[142]

Improved assembly techniques were also introduced at A.V. Roe following delays with the Manchester. The staff felt that 'unless some far better method of feeding the production jigs with detail parts and pre-fabricated sub-assemblies could be devised, the waste of time and labour would be intolerable. And so the idea of kit

marshalling was born.' It was decided to reorganise the storage
of details and any sub-assemblies for each stage in its own kit
marshalling store, to be located adjacent to the appropriate
assembly section. No labour would be allocated to a stage until the
required kit had been marshalled complete in the kit marshalling
store ready for delivery. This represented, at that time, something
of a revolution in organisation. Nevertheless, kit marshalling
'worked smoothly and in a matter of weeks the recorded times were
being returned below target, with great potential savings in jigs and
labour'.[143]

From 1941 the technique spread throughout the industry and
its wider application marked a pronounced departure from the
methods by which production had been accelerated in 1940. The
tendency was, in fact, to rebuild the cushion of parts and components which had been depleted during the Battle of Britain. This
extended the production cycle of individual aircraft, and in peacetime would have been uneconomic in terms of finance (because of
the additional work in progress) and factory capacity (because of the
additional storage space). Yet at a time when the industry was facing
an acute shortage of supervisory staff, kit marshalling reduced
the managerial effort required to maintain the flow of work
and guarded against shortages of particular supplies.[144] Labour
efficiency improved considerably during the war as a result of such
measures (see Table 45). The direct man-hours absorbed in the production of the Lancaster airframe fell from 51,000 in 1941 to
approximately 20,000 by 1945.[145]

TABLE 45
Per Capita Output in the Aircraft Industry, 1941–44

	A) Aircraft Manpower (Million)	B) MAP Production Index Six Months Later	C) Index of Per Capita Output
June 1941	1.2	829	100
Dec. 1941	1.3	1,175	131
June 1942	1.4	1,282	132
Dec. 1942	1.5	1,562	151
June 1943	1.6	1,636	148
Dec. 1943	1.7	1,926	164

Sources: (A) P. Inman, *Labour in the Munitions Industry*, p. 5; (B)
Statistical Digest of the War, p. 154 (MAP production index based on airframe structure weight corrected by man-hours); (C) B ÷ A, June 1941
equated to 100.

THE WARTIME AIRCRAFT INDUSTRY

By the end of 1941, the British aircraft industry was grouped around eight types of aircraft: the three heavy bombers, the Wellington, the Mosquito, the Beaufighter, the Spitfire and the Barracuda torpedo bomber. The administration of these groups was only loosely centralised in the MAP. The parent firm of the most successful group, A.V. Roe, declared in October 1941 that it 'would not be agreeable to operating a Group Central Control if this was to be controlled by the DAP representative', and instead group co-ordination became the responsibility of the leading design firms.[146]

Despite the numerous operational roles which British aircraft had to fulfil, the group system resulted in a high degree of product specialisation. In the airframe sector the principal eight types accounted for 70 per cent of employment by 1944.[147] Of all aero-engines produced in Britain during the war, 81 per cent were designed by just two firms, Rolls-Royce and Bristol, while two models, the Merlin and the Hercules, accounted for 72 per cent of production between September 1939 and May 1945.[148] Propeller production was also dominated by the designs of just two firms, as was turret production.[149] Moreover, the majority of aero-engines and ancillary products were in turn allocated to the most important operational airframes. By June 1944, 50 per cent of total MAP manpower, 850,000 workers, were engaged in producing the Lancaster, the Halifax, and the Wellington alone.[150]

Many other advantages were gained from grouping aircraft production. The groups helped to co-ordinate the supply of components and equipment, and to rationalise tool and material procurement, machining operations, and subcontracting.[151] The solution to maldistribution, for example, lay in setting up group committees to redistribute parts. In November 1941, Rootes received 188 items from other Halifax bomber constructors, while they themselves supplied 35.[152]

Group production also allowed ideas and expertise to be pooled. When Rootes joined the Halifax group, the firm was able to advise Handley Page of several means by which efficiency could be improved. These included the introduction of a new system of factory control, the development of spot welding, and a reorganisation of Halifax assembly which reduced the production time by 20 per cent.[153] In the Lancaster group, A.V. Roe and Metropolitan-Vickers undertook joint research into the application of welding to light alloys.[154] And in the Spitfire group, Westland co-operated with Supermarine in the development of folding wings for the Seafire naval fighter.[155]

Within each group there emerged a two-tier structure of design and production factories. The introduction of step-by-step refinements, which tended to upset manufacturing operations, was undertaken by the principal design firms, but excessive batching of different Marks was avoided at most agency and shadow factories. Supermarine produced several different Marks of Spitfire in parallel for much of the war, but Castle Bromwich rarely produced more than one.[156] Handley Page produced a total of seven different variants of the Halifax bomber between 1941 and 1945, while English Electric produced just three in much larger numbers.[157]

Aero-engine production was planned in the same way. In keeping with pre-war assumptions, Rolls-Royce's Derby and Crewe factories developed and produced the latest versions of the Merlin engine, and later the Griffon, while the Glasgow and Manchester works concentrated on a few proven designs. Upwards of 40 different Marks of Merlin were produced at Derby during the war. Ford, on the other hand, produced just five (very similar) Merlin variants. Yet Ford produced only 2000 fewer engines in five years than Derby produced in seven.[158]

The burden of design and development was therefore borne by the experimental and technical departments of the professional aircraft firms, and high costs were accepted in return for high quality. The most successful types were then selected in consultation with the Air Staff, allocated to mass-production factories, and manufactured in large quantities in long and relatively stable production runs which brought very significant economies. By 1944, the manpower required per Spitfire per month at Castle Bromwich was calculated to be half that required at Supermarine.[159] Similarly Merlin engines produced by Ford cost 30 per cent less than the average cost of those produced by all other Merlin manufacturers.[160]

The conditions under which design and production firms operated differed significantly. Rolls-Royce and Bristol encountered enormous difficulties in co-ordinating engine development with MAP production programmes. The introduction of the two-stage Merlin caused particularly serious problems and, at times, the supply failed to keep pace with installation requirements.[161] As Sidgreaves wrote in 1943, 'our enthusiasm and zeal to introduce improved types has been greater than our real ability to do so'.[162]

De Havilland had eventually to produce no fewer than eight separate variants of the Mosquito at the same time on three 16-stage assembly lines.[163]

Shadow production was, in comparison, relatively simple. At Ford, concentration on individual Marks of the single-stage Merlin increased the scope for scientific factory organisation and for the fulfilment of production targets. Work was moved continuously through a series of bays, each of which specialised in a single process, and operations were carefully subdivided into simple, repetitive, tasks.[164] Standard Motors persuaded the MAP to abandon plans to build two different Mosquito variants at their Canley factory, and production was focused on a single Mark instead. 'This has obvious advantages to us from the production and financial standpoint', the managing director told his board. 'There will be no break in production flow.'[165]

Similar advantages were enjoyed by the Hercules engine shadow group, where long production runs on single Marks of engine allowed very high levels of output to be achieved. Indeed, it seemed to some contemporaries that this type of organisation might hold the key to a more general improvement in productivity. A Ministry of Labour report declared that:

This group of four factories constitutes undoubtedly the best monument and most outstanding achievement of what can be done by careful government planning and methodical industrial execution. Each individual works . . . represents the highest attainment of the engineer's art in mass production methods and they are an object lesson to the rest of the country and indeed to the whole world.[166]

Had all British aircraft production been organised in this way, total output would certainly have been much greater than that actually achieved; but the fundamental strength of the aircraft industry – its flexibility, and its capacity to improve established designs – would then have been wasted, and the quality of aircraft would have deteriorated. In fact, the group organisation established between 1938 and 1941 achieved a balance which maintained both quantity and quality throughout the Second World War.

CONCLUSION

On the outbreak of war the Air Ministry committed itself to a particularly ambitious aircraft programme which initially proved impossible to achieve. Output was expanding rapidly by the spring of 1940 but the plans only really came to fruition after the Air Ministry's responsibilities for aircraft procurement had been transferred to the MAP in May. In the following months the aircraft programme was redrawn; emergency measures implemented by the new administration prolonged the operational life of several established types; capacity had to be created for new designs like the Mosquito, and replacements had to be found for aircraft like the Manchester and the Typhoon.

The 'production miracle' of 1940 encouraged the MAP to plan for even higher targets in the following year but these expectations were soon disappointed. Bombing, dispersal, reduced working hours, design changes and delays in the introduction of new capacity all combined to reduce output. Nevertheless, British aircraft production was placed on a much more stable wartime footing during 1941. The Aircraft Supply Council, new programme and material directorates were created in the MAP. New industrial capacity was sanctioned to improve the balance between the different sectors of the industry and between manufacturing and assembly facilities.

Several firms were reorganised, and there were significant advances in the simplification and standardisation of design and in manufacturing and assembly techniques. By the time the bomber programme was approved in December 1941 a group production system had been established which enabled Britain to concentrate resources and exploit mass production methods without sacrificing the qualitative progress so vital to the operational efficiency of the RAF.

NOTES

1. Postan, *War Production*, pp. 68–9. The Department of Development and Production had been evacuated to Harrogate on the outbreak of war.
2. AC 70/10/56, Handley Page to Bruce-Gardner, 20 October 1939; Handley Page to Fairey, 11 November 1939.
3. AC 70/10/56, Handley Page to Bruce-Gardner, 20 October 1939; Sidgreaves to Handley Page, 21 October 1939; Gouge (Short Brothers) to Handley Page, 21 October 1939; Fairey to Bruce-Gardner, 23 October 1939.
4. AC 70/10/56, Fairey to Handley Page, 13 November 1939.
5. AC 70/10/56, Fairey to Handley Page, 9 November 1939; Handley Page to Fairey, 11 November 1939.
6. AC 70/10/56, Handley Page to Bruce-Gardner, Sigrist (Hawker Siddeley), Fairey, Sidgreaves, Dunbar, and Gouge, 3 November 1939.
7. AC 70/10/56, Handley Page to Bruce-Gardner, 6 November 1939.

8. AC 70/10/56, Fairey to Handley Page, 13 November 1939.
9. PRO AIR 6/35, S of S EPM, 7 September 1938.
10. PRO AIR 6/58, memorandum by DGP, 12 July 1939.
11. PRO AVIA 10/310, DGP meeting, 31 July 1939, 14 August 1939.
12. Vickers 410, Dunbar to AMDP, 18 September 1939; Supermarine daily works report, 29 September 1939.
13. PRO AVIA 10/310, DGP meeting, 18 January 1940; PRO AVIA 10/168, ACCS meeting, 5 February 1940.
14. PRO AIR 6/58, memorandum by DGP, 12 July 1939.
15. PRO CAB 102/275, table by D Stats. P, 9 April 1940.
16. PRO AVIA 10/310, DGP meetings, 18 January, 8 February 1940.
17. Vickers 194, Vickers-Armstrong (Weybridge) quarterly report to September 1939.
18. PRO AIR 6/58, Air Ministry meeting, 12 December 1939.
19. PRO AVIA 10/169, note by DDGP1, 12 February 1940; Vickers 410, Vickers-Armstrong (Southampton) daily reports, 12 and 26 February 1940; Vickers-Armstrong (Weybridge) daily reports, 3 and 19 February 1940.
20. PRO AVIA 10/310, DGP meeting, 2 October 1939.
21. PRO AIR 6/58, Air Ministry meeting, 12 December 1939; PRO AVIA 46/72, DGP to Kingsley Wood, 28 March 1940; Air Ministry memorandum for the Select Committee on National Expenditure, 28 March 1940.
22. Thoms, *War, Industry, and Society*, p. 22.
23. PRO AVIA 46/93, Bruce-Gardner to DGP, 26 January 1940.
24. M. Gilbert, *Winston S. Churchill VI, 1939–41* (London, 1983), pp. 134–5, 192; PRO AIR 19/162, Churchill to Sinclair, 3 June 1940.
25. Scott and Hughes, *Administration*, pp. 291–4, 415–24, 457–65.
26. Postan, Hay and Scott, *Design and Development*, pp. 126–7.
27. PRO AVIA 10/40, unsigned memorandum to S of S, 22 April 1940.
28. Short was nationalised in 1943 and the Stirling was transferred to secondary and transport duties. As late as May 1944, 9 per cent of airframe labour was still employed on the Stirling. See PRO AVIA 10/269, statement of labour statistics prepared by DDGPS, 8 June 1944.
29. Postan, Hay and Scott, *Design and Development*, pp. 134–5.
30. Examples include the Blackburn Skua and Botha, and the Fairey Albacore and Fulmar.
31. Barnet, *Audit*, pp. 146–7.
32. PRO AIR 6/37, S of S EPM, 24 January 1939.
33. M. Donne, *Pioneers of the Skies: A History of Short Brothers Plc* (Belfast, 1987), p. 97.
34. Postan, Hay and Scott, *Design and Development*, pp. 133–8.
35. Ibid., p. 88. The best example is the Spitfire.
36. The aircraft designed to the Air Staff's specification for such a fighter, the Westland Whirlwind, was cancelled in 1940. See PRO AVIA 46/108, Official Historian's type biography of the Beaufighter.
37. PRO AVIA 10/95, DGRD to Hennessy (MAP Controller of Production), 19 February 1941; extract from joint development and production committee, 2 April 1941; PRO AVIA 15/244, DAP to Bristol and Fairey, 26 December 1939.
38. Postan, Hay and Scott, *Design and Development*, p. 86; PRO AVIA 10/188, note by DGAP, 12 October 1941.
39. RRA, Hives file, 'Policy', Hives to Rolls-Royce board, 4 October 1937.
40. AC 70/10/13, Handley Page to Volkert, 23 July 1937.
41. Interview with S.D. Davies, 21 April 1990; 'Personal history of S.D. Davies at A.V. Roe (Manchester), 1938–44'; Connolly papers, 'Survey of the Relationship between Labour Effort, Aircraft Size, Jig and Tool Cost, and Rate of Output', by J.V. Connolly, 15 January 1940; PRO AVIA 46/115, Chadwick to Deputy Director of Research and Development (Airframes), 5 January 1941.
42. ROBN 2/1/3, Pollard (Bristol) to PTCO, 3 July, 1941; PRO AVIA 15/245, meeting of the Beaufighter works managers' committee, 5 May 1941.
43. PRO AVIA 46/116, note by De Havilland, 11 December 1941; Official Historian's type biography of the Mosquito.
44. Postan, Hay and Scott, *Design and Development*, p. 149; interview with S.D. Davies, 21 April 1990.

45. PRO AVIA 15/69, Dunbar to Hennessy, 12 November 1940.
46. Sir Stanley Hooker (assisted by B. Gunston), *Not much of an Engineer: An Auto-biography* (Shrewsbury, 1984), pp. 44–56.
47. Harvey-Bailey, *Hives*, pp. 44–6.
48. PRO CAB 102/50, section 132.
49. PRO AVIA 10/219, ACCS to Vickers-Armstrong (Southampton), 28 May 1940; Harvey-Bailey, *Hives*, p. 44.
50. PRO AVIA 15/874, AD/RDL to AD/RDT1, 11 December 1940.
51. Ibid.; Postan, Hay and Scott, *Design and Development*, p. 533.
52. PRO AVIA 15/874, AD/RDL to AD/RDT1, 11 December 1940; Harvey-Bailey, *Hives*, pp. 45–7.
53. PRO CAB 102/50, 143. Ironically the two-stage Merlin was developed not for the Spitfire but for a high-altitude version of the Wellington. See Hooker, *Not Much of an Engineer*, pp. 52–6.
54. Overy, *Air War*, p. 199.
55. Diagrams showing the organisation of the Ministry are located in the Air Historical Branch archive.
56. Westbrook was, for example, involved in the reorganisation at Bristol early in 1941. See PRO AVIA 15/95.
57. PRO AVIA 10/232, Air Supply Council meeting, 24 May 1941.
58. Postan, *War Production*, pp. 137–8.
59. Ibid., pp. 465–6.
60. Cairncross, *Planning in Wartime*, pp. 10–11.
61. Vickers 389, commercial manager (Chester) to H.L. Wonfor (Vickers-Armstrong) 21 March 1941.
62. RRA, Hives file, 'Hennessy', Hives to Hennessy, 2 January 1941.
63. PRO CAB 102/51, section 117; PRO AVIA 10/311, *UK New Engine Deliveries by Firms, 1939–45*.
64. PRO AVIA 15/3677, PAS (E) to PS2, 3 September 1941; memorandum by Permanent Secretary, 18 September 1941.
65. PRO AVIA 15/3677, DDG Stats. P to Cherry (MAP personnel branch) 4 August 1942.
66. PRO AVIA 10/378, note by DDG Stats. P, 12 February 1942.
67. Cairncross, *Planning in Wartime*, pp. 30, 64.
68. PRO AVIA 15/29, AMDP to CAS, 12 May 1940.
69. PRO AVIA 15/29, CMDP to PS4, 15 May 1940.
70. Vickers 410, Vickers-Armstrong (Weybridge) daily works report, 9 December 1939.
71. The percentages for nine types were: Hurricane 16, Spitfire 3.5, Blenheim 8, Hampden 6, Wellington 2, Whitley 1, Lysander 5, Oxford 3, Anson 7. See PRO AVIA 10/181, table by Director General of Equipment, 13 February 1941.
72. PRO AVIA 15/1839, memorandum by the sub-committee of the Aircraft Subcontracting Committee, 22 May 1943.
73. PRO AVIA 10/176, note by Director General of Aircraft Production Factories, 2 July 1940.
74. PRO AVIA 10/176, note by DEP, 15 August 1940.
75. PRO AVIA 46/73, memorandum by Assistant Director of Materials Production (5), 12 August 1943.
76. PRO AVIA 10/219, Dunbar to AMDP, 23 November 1939.
77. PRO AVIA 10/219, ACCS meeting, 30 November 1939.
78. Vickers 198, Vickers-Armstrong (Southampton) quarterly report to September 1940.
79. Vickers R26, company secretary to US of S, Air Ministry, 30 May 1940.
80. Vickers 416, Dunbar to Joseph Lucas Ltd, 21 July 1940.
81. Vickers 410, Dunbar to Craven, 8 March 1940; Vickers 416, Dunbar to Yapp, 6 August 1940.
82. Vickers 416, Anderson to Palmer, 20 July 1940.
83. Vickers 416, Dunbar to Yapp, 14 July 1940.
84. PRO AVIA 46/168, 'The Repair and Maintenance of Aircraft', by E. Bridge (unpublished official narrative), p. 61.
85. Overy, *Air War*, pp. 42, 192.
86. PRO AIR 6/58, memorandum by DGP, charts 5 and 9, 12 July 1939.

87. Vickers 199, Vickers-Armstrong (Southampton) quarterly report to December 1940.
88. Vickers 198, 199, and 200, Vickers-Armstrong (Southampton) quarterly reports to September and December 1940, and to March 1941.
89. For Supermarine production figures see PRO AVIA 10/311, *UK New Aircraft Deliveries by Firms.*
90. Vickers 198, Vickers-Armstrong (Weybridge) quarterly report to September 1940.
91. Vickers 416, Dunbar to Craven, 20 August 1940.
92. PRO AVIA 10/267, table by DDG Stats. P, 20 December 1941.
93. Mensforth, 'Airframe Production', p. 37; PRO LAB 8/614, report by the Production Efficiency Board of the MAP, 12 August 1943; Vickers 197, Vickers-Armstrong (Weybridge) quarterly report to June 1940; MAP Monthly Statistical Bulletin. The productive labour rose from 5313 in July 1940 to 7180 in March 1943; the total labour force from 6731 in June 1940 to 11,211 in June 1943. In Germany the indirect labour force increased by 20 per cent because of dispersal. See USSBS, Report 4, p. 26.
94. See for example PRO CAB 102/53, p. 31.
95. Hornby, *Factories*, p. 205; PRO CAB 16/227, statement by the Air Ministry of expenditure on Air Ministry shadow factories, March 1936–December 1938.
96. PRO AVIA 10/200, note by DGAP and PAS, 18 September 1943; PRO AVIA 15/1839, memorandum by the subcommittee of the Aircraft Subcontracting Committee, 22 May 1943; PRO AVIA 15/1466, DDGAP to Dobson, 28 February 1945; PRO AVIA 15/244, meeting of Beaufighter group committee, 4 January 1940. The policy of duplicating sources of supply was introduced by the Air Ministry during 1938 and 1939.
97. Vickers 200, Vickers-Armstrong (Southampton) quarterly report to March 1941.
98. PRO AVIA 10/170, memorandum by DDGP2, 5 March 1940.
99. PRO AVIA 10/184, ASB meeting, 19 June 1941.
100. RRA, Hives file, 'Sir Charles Craven', Kelley to Hives, 28 November 1941.
101. AC 72/3, 74.
102. PRO AVIA 15/245, meeting of Beaufighter works managers' committee, 5 May 1941.
103. *Statistical Digest*, p. 154. The index was based on deliveries of airframe structure weight and man-hours per airframe equated to 1000 in January 1942.
104. Hornby, *Factories*, p. 211.
105. Ibid., pp. 265–79.
106. PRO AVIA 10/178, report by DMP, 23 November 1940.
107. Postan, *War Production*, pp. 159, 311; Mensforth, 'Airframe Production', p. 30; PRO AVIA 46/73, AD/MP5 to Postan, 12 August 1943.
108. PRO AVIA 10/108, E. Walton to T.P. Wright, 24 July 1943.
109. Postan, *War Production*, p. 203; Hornby, *Factories*, p. 211.
110. On labour-saving investment see PRO AVIA 10/197, note by DGAP, 8 December 1942; PRO AVIA 10/198, note by DGAP, 15 February 1943; PRO AVIA 10/199, note by DGAP, 22 July 1943.
111. PRO AVIA 15/2660, report by North American Aviation on weight comparison between Mustang and Spitfire, February 1943; PRO AVIA 10/100, Consolidated report on Inspection Trip of US Aircraft Production Mission to England, 14 October to 11 November 1942, pp. 26, 51–3.
112. Postan, *War Production*, p. 321.
113. PRO AVIA 10/269, statement of labour statistics prepared by DDGPS, 8 June 1944.
114. PRO AVIA 15/1434, Churchill to the Minister of Aircraft Production, 12 July 1941; MAP meeting, 17 July 1941.
115. Hornby, *Factories*, pp. 209–10.
116. Hornby, *Factories*, pp. 216–7.
117. Lloyd, *The Merlin at War*, p. 77.
118. ROBN 2/1/3, PTCO to the Minister, 3 July 1941; Pollard to PTCO, 3 July 1941.
119. Austin archive, MSS 226/AU/1/1/2, report by Mr Lord to the Austin board, 29 January 1941.
120. HSA, Hawker board minutes, 29 January 1941, 17 February 1941; PRO AVIA 10/183, memorandum by DGAP, 11 May 1941.
121. Ibid.
122. PRO AVIA 10/183, note by DGAP, 16 March 1941.

123. PRO AVIA 10/186, memorandum by DGAP, 20 August 1941; PRO AVIA 10/187, ASB meeting, 16 September 1941.
124. PRO CAB 102/187, pp. 19–36.
125. Postan, *War Production*, pp. 316–18.
126. See Chapter 3.
127. PRO AVIA 10/186, note by DGAP, 30 July 1941; PRO AVIA 10/189, note by DGAP, 17 November 1941.
128. Ibid.
129. Mensforth, 'Airframe Production', pp. 31–2.
130. PRO AVIA 10/198, note by DGAP, 30 March 1943.
131. PRO AVIA 46/105, Postan's interview with Chadwick, 6 January 1944; Thoms, *War, Industry, and Society*, p. 43.
132. PRO AVIA 10/195, note by DGAP, 25 June 1942.
133. Mensforth, 'Airframe Production', p. 32.
134. PRO AVIA 46/105, Postan's interview with Chadwick, 6 January 1944.
135. Interview with S.D. Davies, 30 April 1990.
136. Mensforth, 'Airframe Production', p. 32.
137. A. Harvey-Bailey, *The Merlin in Perspective: The Combat Years* (Rolls-Royce Heritage Trust, Derby, 1983), p. 32.
138. G. Mills, 'Ford and the Merlin', *The Archive*, Vol. 9, No. 27 (1991), p. 36.
139. AC 70/10/58, memorandum by the SBAC, 8 January 1943.
140. Mensforth, 'Airframe Production', p. 26.
141. PRO AVIA 10/166, memorandum by DGP, 16 December 1939; F.G. Miles, 'Track Assembly', *Flight*, Vol. 38, No. 1652 (August 1940), pp. 146–9.
142. PRO BT 28/423, memorandum by Sir Ernest Lemon, 23 June 1942; Mensforth, 'Airframe Production', p. 33.
143. 'Personal history of S.D. Davies at A.V. Roe (Manchester) 1938–44'. In 1939–40 Davies, who was assistant chief designer, was seconded to A.V. Roe's production department to help familiarise production staff with the new Manchester bomber.
144. Mensforth, 'Airframe Production', p. 30; Anon., 'The Control of Planned Production', *Aircraft Production*, April 1941, pp. 121–2; Hornby, *Factories*, pp. 236–8.
145. PRO AVIA 10/268, table by DDG Stats. P, 4 October 1941; PRO CAB 120/351, memorandum by the Minister of Aircraft Production and the Secretary of State for Air, 'Modifications and Aircraft Production', 5 March 1945.
146. PRO AVIA 15/1465, statement by A.V. Roe and Co. Ltd, on a suggested scheme for the formation of central control for the Lancaster group, 5 October 1941; Postan, *War Production*, p. 419.
147. PRO AVIA 10/269, memoranda by DDGPS, 20 May, 8 June 1944.
148. Calculated from PRO AVIA 15/311, UK New Engine Deliveries by Firms, 1939–45; *Statistical Digest*, p. 155; Hornby, *Factories*, p. 254.
149. Hornby, *Factories*, pp. 267, 274–5.
150. PRO AVIA 10/269, memoranda by DDGPS, 20 May 1944, 8 June 1944.
151. Mensforth, 'Airframe Production', p. 26.
152. PRO AVIA 15/1591, ADAP to DGAP, 29 October 1941; Halifax group co-ordination meeting, 11 November 1941.
153. AC 70/10/186, Tyrrel (Rootes) to Ratcliffe (Handley Page), 1 and 3 June 1943.
154. Interview with S.D. Davies, 30 April 1990.
155. Mensforth, *Engineers*, p. 76.
156. MAP *Monthly Statistical Bulletin*.
157. AC 78/23, papers on Halifax group.
158. Information supplied to the author by Rolls-Royce.
159. PRO AVIA 10/269, table by DDGPS, 20 May 1944.
160. PRO CAB 102/51, section 208, note 1.
161. PRO AVIA 10/393, supplements to MAP statistical bulletin by DDG Stats. P, 16 June 1943, 15 December 1943.
162. RRA, Hives file, 'MAP general, 1941–43', Sidgreaves to Lord Herbert Scott, 19 September 1943.
163. PRO AVIA 9/51, Cripps to Sinclair, 16 November 1943.
164. PRO CAB 102/51, section 207–8.

165. Standard Motors archive, ST 1/1/7, managing director's report, 20 July 1943.
166. Standard Motors archive, ST 1/1/7, managing director's report, quoting the Ministry of Labour and National Service, 25 June 1941.

Conclusions

In July 1934, Neville Chamberlain, then Chancellor of the Exchequer, overturned the recommendations of the government's Defence Requirements Committee by imposing drastic cuts on its proposals for army and navy rearmament. At the same time Chamberlain directed that the strength of the Metropolitan Air Force be raised from the 52 squadrons proposed by the committee to 80 squadrons.[1] This symbolic intervention established an order of priorities which was retained throughout the rearmament years. It was based on the belief that the next war would be an air war: a war that would be won or lost in the air.

The government's foresight owed much to conclusions drawn from the First World War. Air power had robbed Britain of centuries of geographical immunity from invasion, but it appeared to offer an alternative to committing large ground forces to protracted trench-warfare on the continent of Europe. Given Britain's traditional aversion to large land armies, and the resulting 'capital intensive' orientation of her defence policy, it is hardly surprising that British thinking on aerial warfare was more advanced than in other countries.

By 1939 the RAF had a far clearer understanding of the defensive capabilities of fast monoplane fighters than did air forces elsewhere; it was the most outspoken advocate of strategic bombing, and it had also developed a clear doctrine on the use of air power in conjunction with land and sea operations. Although the theory had inevitably to be adapted to wartime realities, British strategists accurately predicted the type of war which was to be fought after 1939 and the weapons needed to fight it.[2]

The industrial consequences of Britain's commitment to air power can be expressed in purely numerical terms. In 1935 the British aircraft industry built just 893 aircraft for the RAF. In 1941 it produced more than 20,000, and in 1944 more than 26,000. The figures are even more impressive when expressed in terms of structure weight. The industry delivered less than 2 million pounds in

1935; in 1941 it delivered more than 87 million pounds, and in 1944 more than 208 million pounds. Fewer than 3000 aero-engines were delivered in 1936. In 1941 36,500 engines were built; in 1944 this increased to 57,000 engines. On average each aero-engine produced in 1944 developed double the horsepower of each engine built in 1936. The MAP's production index increased from an average monthly figure of 18 in 1935 to 803 in 1941, and to 1774 in 1944.[3] The expansion of the aircraft industry was achieved by investing more than £350 million in factories and plant for new production between April 1936 and March 1944, and by increasing employment from 35,000 in 1935 to nearly 1.7 million in 1943.[4]

The figures alone, however, tell us nothing about the allocation of resources, about how expansion was planned, or about how the industry responded to the pressures of rearmament and war. This study has attempted to answer these questions. Its principal conclusions may now be summarised.

The demand for military, civil and private aircraft between the wars was sufficient to support a relatively large aircraft industry in Britain. It is true that certain companies had to diversify their activities to maintain design capacity and that some of the most successful firms of the 1920s were unable to hold on to their share of the market after 1930 but this only helped to advance a process of consolidation which resulted in the emergence by 1935 of the key wartime (and postwar) contractors, Vickers, Hawker Siddeley, Bristol, De Havilland and Rolls-Royce. These firms were highly profitable and were able to gain considerable production experience before the commencement of rearmament.

Nevertheless, consecutive Air Ministry studies demonstrated that the industry was not in itself capable of meeting the needs of war production, so plans were drawn up to augment the capacity of the professional airframe and aero-engine firms by employing automobile and other engineering companies as 'shadows', to concentrate industrial resources into groups, and to create a state production organisation to co-ordinate the output of the different sectors of the aircraft industry. It was expected that these plans might have to be implemented without notice to bridge the gap between RAF requirements and industrial capacity in the first 12 months of a war but, in the event, Britain was granted a breathing space of some five years in which to rearm.

A feature of the early rearmament years was the sharp distinction which the government drew between peacetime and wartime pro-

duction planning. From 1934–38 the aims of the defence programme were strictly limited, and these limitations were reflected
in the structure and authority of the state's aircraft production
organisation. Responsibility for design and production was initially
divided between two different Air Ministry departments. A separate
production directorate, the DAP, was formed within the Department of Supply and Organisation at the beginning of 1936. The
DAP was given the function of monitoring the progress of individual contracts, but it had little or no control over the way in which
the industry was organised and its attempts to draw up production
programmes were initially unsuccessful. 'DAP has attempted to
forecast deliveries', Freeman wrote at the end of 1937, 'but to do so
with any accuracy has been impossible since the manufacturers had
no experience of production of the new types.'[5]

When rearmament was accelerated following the *Anschluss*, however, the Air Ministry's aircraft design, development and production branches were rapidly integrated, and greater numbers of
qualified staff were recruited. Long-term war production targets
provided a basis for the Ministry to plan new capacity and to co-
ordinate the output of different aircraft components, equipment and
materials. Programming and co-ordination inevitably suffered
during the emergency of 1940; the Ministry of Aircraft Production
assumed control of air force supply with a mandate to remedy what
was mistakenly seen as a production crisis, and 'Beaverbrook's concern was to enlarge the immediate output of whatever was in production, not to alter existing plans so as to achieve better results
later on.'[6] Nevertheless, progress was resumed in the following year
after the creation of the MAP's Directorate of Statistics and
Programmes.

A cohesive state production policy also developed in stages. In
1935, RAF specifications were relaxed, the Capital Clause was
introduced to encourage investment, two-year contracts were allocated to certain firms, and the progress payment scheme was
extended. In the following year the government approved far more
ambitious plans to re-equip the RAF with modern aircraft. Three-
year contracts were placed, and further financial concessions were
granted to the aircraft industry. The first shadow factories were set
up, although they were intended only to help fulfil the peacetime
rearmament programme.

In 1938 planning for war production began in earnest. The
government started investing directly in the aircraft firms and the

industry agreed to manage new factories built at state expense. The first group production schemes were organised and new shadow factories were established under companies like Nuffield, English Electric, Vickers-Armstrong and Metropolitan Vickers. Subcontracting tapped substantial resources of labour, management, machinery and working capital which might otherwise have been excluded from the aircraft economy.

Throughout the rearmament years the Air Ministry failed to achieve a significant reduction in the number of different aircraft types in service. However, premature concentration might well have eliminated some of the best designs like the Spitfire, the Lancaster and the Mosquito, and by ordering different models to perform the same military functions the Ministry insured itself against the technical failures which proved so disastrous in Germany in 1942. Moreover, the relatively large number of different models produced in Britain disguised enormous variations in the quantity of resources allocated to each one. The organisation of group production schemes for the Wellington, Halifax, and Manchester bombers and for the Merlin and Hercules engines ultimately resulted in a very high degree of product specialisation during the war.

When the aircraft industry was required to expand in 1935, the complex technical transition from the biplane to the monoplane was far from complete. The quality of British aircraft had advanced considerably during the inter-war years but the greatest improvements in performance had been achieved by increasing the ratio of aero-engine power to aircraft weight. The lessons of the aerodynamic revolution associated with large civil land-planes in the United States were rapidly being learned but the first British military monoplanes were still being developed at the beginning of rearmament.

The technical problems associated with the introduction of this new generation of aircraft imposed far more rigid constraints on production than did purely physical shortages of particular industrial resources. Such shortages certainly occurred but it is unlikely that a more rapid influx of manpower or machine tools would have produced significantly greater numbers of modern aircraft in 1937 or 1938, nor would the higher levels of government defence spending demanded by Churchill have accelerated preparations to any material degree. Until the development stage had been completed and production drawings standardised, design capacity represented the single most important obstacle to the fulfilment of the production targets.

Yet the problem initially went unrecognised. Some firms were overburdened with design work, and new types were scheduled for introduction before they were fully developed. Freeman recalled later that

[w]e became opportunists and took or ordered what we thought we could get, believing in the facile promise of the manufacturers who, in order to make certain of getting orders, made unduly optimistic promises of what they could produce. We cannot altogether blame them for they were in some measure encouraged to do so by our urging them on to increase their promises if they fell short of our wishes.[7]

The Air Ministry was, in fact, aiming very high indeed. Despite a substantial increase in output and labour productivity during 1937, most of the more important production deadlines were missed. Believing that aircraft which had been ordered at the beginning of rearmament would be obsolete by the time they entered service, the Ministry leaned hard on the firms who, fearful of compulsion or outright nationalisation, attempted to shift responsibility for their troubles back on to the government. In the aero-engine industry the problems were less severe but the task of planning was complicated by the slow progress of the airframe programme and by the prevailing uncertainty about the government's long-term intentions, and some time passed before the industry began to fulfil its true potential. During 1938, however, the position improved. The future demand for military aircraft was clarified when scheme L was finally sanctioned by the government; through the medium of the Supply Committee, state and industry began to co-operate much more closely in planning new capacity; most importantly of all, the new monoplanes finally began to emerge in quantity.

By the end of the year, levels of output considerably higher than originally expected were allowing at least some of the accumulated deficit to be made up. During 1939 British aircraft production overtook German aircraft production. In 1940 Britain produced more aircraft than any other country in the world, and in 1941 Britain outproduced Germany by 71 per cent in terms of complete aircraft, by 28 per cent in terms of structure weight, and by 63 per cent in terms of aero-engines, despite employing fewer resources of labour and aluminium. Why did production accelerate so rapidly?

The rate of expansion is partly explained by the very high level of investment in new factories, plant and machinery undertaken by the aircraft industry between 1935 and 1938. Until the technical constraints on production had been overcome this capacity remained underutilised; it only fulfilled its true potential in 1939. State investment then accelerated the production of types like the Wellington, Hurricane and Spitfire during 1939 and 1940 and added to the amount of floor space available for the heavy bombers which were to be introduced under the auspices of the War Potential programme in 1940 and 1941. 'Concerted and vigorous action' was taken to enlarge capacity for ancillary products like aero-engines, aluminium, gun turrets and equipment.[8]

These measures provided the foundation for the high production levels achieved during the Battle of Britain.[9] Investment by the MAP was chiefly directed towards adapting industrial capacity to the longer-term demands of the wartime air force: to unpredicted and largely unpredictable changes in the design and development programme, and to the introduction of the bomber programme at the end of 1941.

The expansion of British aircraft production could not have been sustained without adequate supplies of manpower. An acute shortage of draughtsmen and technicians was the root cause of many of the industry's problems in the early years of rearmament. By 1938, however, growing awareness of this shortage was encouraging both state and industry to concentrate design effort on the most important short-term projects, and the creation of self-contained experimental departments by several leading contractors helped to accelerate the development of such key aircraft as the Mosquito and the Lancaster. Similar bottlenecks occurred in the supply of supervisory personnel, especially after the dispersal of the industry at the end of 1940, but shadow production and subcontracting helped to spread the managerial burden that wartime expansion imposed on the aircraft firms, and the data for production and resource utilisation do not indicate that Britain was any more severely handicapped by this problem than Germany.

In contrast, the aircraft industry could always draw on relatively liberal supplies of labour. There was initially some entrenched union opposition to dilution but it was impossible for the unions to prevent new factories from being staffed with unskilled workers. Union resistance became futile after 1938 as more and more new capacity was sanctioned for the War Potential programme, hence the remarkable contrast between Blackburn's problems at Dumbarton and Rolls-Royce's success in recruiting an almost entirely unskilled work-force in Glasgow.

The massive influx of unskilled workers created its own

problems. Absorption took time; it depended on the transfer of a nucleus of skilled workers from the older factories and it also required elaborate jigging and tooling. The higher degree of mechanisation, employed during production runs of unexpected length and stability, resulted in dramatic improvements in per capita output. The time taken to build the centre section of a particular bomber fell from 1014 hours at the commencement of production to 592 hours after one year, and to 230 hours after 17 months.[10] Output benefited enormously, but accurate rate fixing became virtually impossible. Excessive earnings were, however, mainly confined to the Midlands shadow factories, and the available figures do not suggest that high piecework bonuses prevented labour productivity from increasing throughout the war.

Rising investment and employment fuelled the expansion of British aircraft production between 1939 and 1941; but the growing volume of turnover had also to be financed. During the aircraft boom of the mid-1930s the nominal value of publicly issued capital in the aircraft industry increased from £3.7 million to more than £19 million, but as production began to outstrip the industry's own financial capacity the state's role in providing capital inevitably increased. Very large numbers of aircraft were built under agency agreements after 1938, but the government preferred to maintain operations on a commercial basis wherever possible, and state capital assistance schemes and bank-lending provisions were therefore liberalised. When it became clear that the firms were also employing excess profits to finance production the government turned a blind eye. Yet this did not mean underwriting exorbitant production expenses. The pricing mechanism thrashed out between state and industry during the rearmament years rewarded firms for the improvements in efficiency which the government was so anxious to achieve and imposed effective long-term controls over aircraft costs.

Increasing efficiency reflected wide-ranging managerial reform, as in the case of Rolls-Royce, and continuous technical innovation. The greatest economies resulted from the simplification of design, from extensive jigging and tooling, from splitting up complete aircraft into large numbers of component sub-assemblies, from the improvement of metal-forming techniques, and from the increasingly widespread use of progressive assembly.[11] Flow production methods and single-purpose machinery also allowed the aero-engine industry to achieve significant gains in productivity.

Historians, however, have invariably judged the aircraft
industry's wartime performance not by any detailed assessment of
its productivity record but by its failure to fulfil the official production programmes, despite its apparently endless consumption of
human and capital resources. A former programmer recently
blamed this failure on 'sheer inefficiency'.[12]

This is an excessively harsh verdict. There is no quantitative
evidence that the British aircraft industry was less efficient than its
German counterpart. British output, expressed in terms of the
structure weight of complete aircraft, was significantly higher than
German output between 1941 and 1944. In 1942 Britain produced
more than 50 per cent more aircraft than Germany and nearly 40
per cent more airframe structure weight. In 1943 Britain again out-
produced Germany in absolute terms, and in 1944 Britain still produced nearly 20 per cent more airframe structure weight than
Germany (see Table 46), yet the German aircraft industry was
employing more labour by 1941 than the British aircraft industry
employed throughout the war.[13]

TABLE 46

British and German Aircraft Production, 1941–44

Year	Structure Weight (m. lb) Delivered		Aircraft Delivered	
	Britain	Germany	Britain	Germany
1941	87	68	20,094	11,776
1942	133	96	23,672	15,409
1943	185	142	26,263	24,807
1944	208	175	26,461	39,807

Source: *Statistical Digest of the War*, p.153; USSBS, Report 4, exhibit
VI A; R.J. Overy, *The Air War*, p. 192.

Per capita output in Britain was lower than in America. The most
reliable comparison between the American and British airframe
sectors concluded that output was about 75 per cent higher in the
former than in the latter. But behind the American advantage lay
production targets many times greater than those set in Britain. The
scale of the American aircraft programme provided the justification
for a level of fixed investment which dwarfed that undertaken by the
European powers, and lower production costs also reflected the
greater volume of production in the United States and the very

much larger average size of American factories. An equivalent degree of industrial concentration was, of course, impossible in Britain because of the threat of German bombing.[14]

There were many reasons for the failure of British aircraft production to fulfil the programme targets during the war. By 1941 the industry was dependent on a massive international manufacturing effort. Bauxite produced in the colonies was turned into aluminium in Canada, before its conversion into alloy in Britain. America supplied machine tools and many other products needed to produce finished aircraft. The level of mobilisation and output achieved in Britain would have been impossible without the support of the Empire and the United States but bottlenecks and production shortfalls could also result from circumstances entirely beyond British control. Nevertheless, the real explanation for the gap between programme and production lay in the frequency with which aircraft were modified in accordance with service requirements. The solution – the imposition of stricter limitations on programme changes and modifications – lay chiefly with the government and the RAF rather than the aircraft industry, for improved manufacturing methods could not have been exploited in the absence of long production runs on standardised products, as the government's own research showed.[15] In the later stages of the war both Germany and America made substantial qualitative sacrifices in the interests of maximising output.[16] Quality was also sacrificed to quantity at mass-production factories in Britain like Castle Bromwich or Ford's Merlin engine plant at Manchester. At design factories such as Rolls-Royce and Supermarine, however, the production schedules were repeatedly upset by the continuous process of qualitative refinement. Some loss of output was expected, but there was no scientific basis for calculating how great that loss would be.

The official programme can therefore hardly be seen as the only yardstick of the aircraft industry's productive efficiency and there are indeed several other indicators, none of which show that resources were inefficiently employed. There is no evidence of extravagance in the supply of machine tools; on the contrary, capital equipment in the British war economy as a whole was always much more fully utilised than in Germany.[17] The bulk of aluminium scrap was ploughed back into finished aircraft; prices remained stable despite the increasing weight and complexity of aircraft and despite the steady rise in wage rates; subcontracting did limit productivity, as it did in Germany, but it also greatly reduced the need for direct government investment; and, even after the production figures were corrected to account for changes in aircraft weight, they still revealed a steady increase in per capita output between 1941 and 1944. This conclusion is corroborated by the MAP's own calculation, in 1944, that per capita output in the airframe sector alone had increased between three and four times since 1936.[18] This was achieved despite major quantitative sacrifices during 1943 and 1944 which were justified by the Ministry on the grounds that the quality of aircraft would improve.[19] The programmers were indulging in excessive optimism if they truly expected to obtain much more.

After 1945, the history of British aircraft production in the Second World War became shrouded in a popular mythology which dwelt on individual designers of genius, like Mitchell and Camm, on their aircraft, the Spitfire and the Hurricane, on 'boys in the backroom' and on supposed miracle workers like Lord Beaverbrook. More recently historians have shown how easily the myth can be debunked. In its place there now stands a gloomy catalogue of muddles and missed delivery dates which is compared unfavourably to an idealised (but very misleading) evaluation of German and American aircraft production.

In fact the achievements of the British aircraft economy in the Second World War did not stem purely from the work of a handful of brilliant engineers and administrators. Rather, they were the result of consistent and methodical plans for wartime production first proposed in the 1920s and finally implemented during the rearmament and early wartime years, and of sustained effort and innovation on the part of the leading firms. Indeed, this was one of the very few instances in Britain's industrial history when it has been recognised that successful manufacturing operations require careful long-term planning and preparation, and close co-operation between the state and private enterprise.

It is true that the plans did not always work out as expected, chiefly because the impact of rapid technological change on production was impossible to predict. Yet by the end of 1941, as the War Potential programme came to an end and as the new bomber programme was launched, the aircraft industry had been organised into a number of groups geared to the development and production of the most important airframes and ancillary products. The groups' effectiveness lay in their productive capacity and in their ability to respond flexibly to changing service requirements. The maintenance of this balance between the quantity and the quality of military aircraft – notoriously more complex and technologically advanced than any other product of the armaments industry – was the single most outstanding accomplishment of the British war economy.

NOTES

1. J. Terraine, *The Right of the Line* (London, 1988), pp. 25–30.
2. Overy, *Air War*, Chapters 1 and 9.
3. *Statistical Digest*, pp. 151–5.
4. Hornby, *Factories*, pp. 212, 251. For new aircraft and spares total employment in December 1943 amounted to 1.5 million.
5. PRO AIR 6/51, note by AMRD, 3 December 1937.
6. Cairncross, 'Reflections: How British Aircraft Production Was Planned in the Second World War', *Twentieth Century British History*, Vol. 2, No. 3 (1991), p. 347.
7. PRO AIR 6/51, note by AMRD, 3 December 1937.
8. PRO AIR 6/58, memorandum by DGP, 12 July 1939.
9. A.J. Robertson, 'Lord Beaverbrook and the Supply of Aircraft, 1940–1941', in A. Slaven and D.H. Aldcroft (eds), *Business, Banking and Urban History: Essays in Honour of S.G. Checkland* (Edinburgh, 1982).
10. Mensforth, 'Airframe Production', p. 36.
11. Ibid., p. 24.
12. Cairncross, *Planning in Wartime*, p. 64.
13. Overy, *Goering*, p. 176.
14. Mensforth, 'Airframe Production', p. 36. Mensforth employed a somewhat different measurement of airframe weight and made allowances for the production of spare parts.
15. PRO BT 28/423, memorandum by Sir Ernest Lemon, 12 August 1942.
16. Overy, *Air War*, pp. 228–30.
17. USSBS, Report 54, *Machine Tools and Machinery as Capital Equipment*, pp 13–14 'Germany's machine tool equipment was much less fully utilised than Great Britain's'.
18. PRO BT 28/161, MAP press release on aircraft production figures, June 1944.
19. PRO CAB 92/40, memorandum by the MAP for the Joint War Production Staff, 19 June 1943.

Bibliography

MANUSCRIPT AND ORIGINAL SOURCES

Records in public depositories

Air Historical Branch Archive, Ministry of Defence
Ministry of Aircraft Production Monthly Statistical Bulletin.

Cambridge University, Churchill College:
Lord Weir papers
    Accession numbers 19/1, 2, 5, 13, 16, 20, 21, 23, 23A.
Professor Sir Austin Robinson papers:
    Accession number 2/1/3.

Cambridge University Library
Vickers archive:
    Accession numbers K757, R26, 61, 163, 169, 185, 188, 189, 194,
    196, 197, 198, 199, 200, 201, 206, 212, 222, 315, 322, 324, 326,
    339, 369, 389, 410, 416, 626, 678, 687, 701, 780, 866, 1245, 1371,
    1392, 1798, 1799.

Coventry City Record Office
Armstrong Siddeley archive:
    Accession number 1060/1/3.
Alfred Herbert archive:
    Accession number 586/30/1.

Public Record Office, Kew
AIR 2. Air Ministry registered files:
    Piece numbers 266 (S.23447), 619, 714, 1208, 1213, 1322, 1668,
    1790, 1908, 1951, 2014, 2572, 2577, 2711, 3273, 9165, 9167, 9168.
AIR 6. Records of meetings of Air Board and Air Council:
    Piece numbers 23, 24, 26, 30, 31, 32, 33, 34, 35, 36, 37, 43, 45, 47,
    48, 49, 51, 52, 53, 54, 55, 58.

AIR 8. Chief of the Air Staff files:
Piece number 196.
AIR 19. Air Ministry Private Office papers:
Piece numbers 2, 4, 5, 9, 162, 524.
AIR 20. Air Ministry unregistered papers:
Piece number 2379.
AVIA 8. Air Ministry Research and Development files:
Piece number 158.
AVIA 9. Ministry of Aircraft Production Private Office papers:
Piece numbers 51, 70.
AVIA 10. Ministry of Aircraft Production unregistered papers:
Piece numbers 6, 9, 12, 40, 95, 100, 102, 108, 151, 153, 154, 155,
156, 165, 166, 167, 168, 169, 170, 176, 178, 183, 184, 186, 187,
188, 189, 195, 197, 198, 199, 200, 217, 219, 241, 242, 246, 253,
232, 241, 243, 245, 247, 253, 267, 268, 269, 310, 311, 357, 378,
393.
AVIA 15. Ministry of Aircraft Production files:
Piece numbers 29, 69, 95, 159, 244, 245, 268, 311, 639, 874, 1040,
1434, 1465, 1466, 1591, 1839, 2346, 2540, 2660, 3677, 3710, 3718.
AVIA 46. Ministry of Supply files series I (Establishment):
Piece numbers 31, 35, 72, 73, 88, 91, 93, 105, 108, 110, 112, 113,
114, 115, 116, 168, 210, 268, 498.
BT 28. Ministry of Production correspondence and papers:
Piece number 423.
CAB 4. Defence Policy and Requirements Committee:
Piece number 4.
CAB 16. Committee of Imperial Defence ad hoc sub-committee
papers:
Piece number 227.
CAB 21. Cabinet Office registered files:
Piece number 703.
CAB 60. Principal Supply Officers Committee Supply Board
memoranda:
Piece number 34.
CAB 87. War Cabinet committees on reconstruction:
Piece number 54.
CAB 92. War Cabinet committees on supply, production, priority
and manpower:
Piece number 40.
CAB 102. Cabinet Office historical section: official war histories
(Second World War), civil:
Piece numbers 40, 47, 50, 51, 53, 187, 239, 240, 274, 275, 511.

CAB 120. Minister of Defence: Secretariat files:
Piece number 351.
ED 46. Records created or inherited by the Department of Education and Science, Further Education, general files:
Piece numbers 226, 257.
LAB 8. Ministry of Labour, employment files:
Piece numbers 217, 374, 614.
LAB 76. Ministry of Labour official histories, correspondence and papers:
Piece number 28.
SUPP 3. Records of the Principal Supply Officers Committee of the Committee of Imperial Defence:
Piece numbers 9, 44.
T 161. Treasury supply files:
Piece numbers 922/S.40730/04, 1316/S.40700/7-63, 1323/S.40700/42, 1325/S.40700/56.

Royal Air Force Museum, Hendon
Bristol archive:
Accession numbers AC 79/2, AC 72/3.
Sir Christopher Bullock papers (no accession numbers).
Handley Page archive:
Accession numbers AC 70/10/9, AC 70/10/13, AC 70/10/54, AC 70/10/39, AC 70/10/55, AC 70/10/56, AC 70/10/58, AC 70/10/67, AC 70/10/79, AC 70/10/139, AC 70/10/159, AC 70/10/169, AC 70/10/186, AC 78/23, HP 8, HP 759, works order cost records (no accession numbers).
Ministry of Aircraft Production price books (no accession numbers).

University of Warwick Modern Record Centre
Austin Motors archive:
Accession number MSS 226/AU/1/1/2.
Engineering Employers' Federation archive (including minutes of the Aircraft Manufacturers' National Technical Committee and labour returns):
Accession numbers MSS 237/1/6 and MSS 237/13/3.
Federation of British Industries archive:
Accession numbers MSS 200/F/3/52/15/16 and 34.
Standard Motors archive:
Accession number MSS 226/ST 1/1/7.

Rover archive:
  Accession number MSS 226/RO/1/4/1.

Records held privately

Hawker Siddeley, Slough
Hawker Siddeley archive: Airspeed Policy Committee minute book, Armstrong Whitworth board minutes and accounts, Blackburn board minutes, Blackburn Consolidated board minutes, De Havilland board minutes and accounts, Folland Aircraft board minutes, Hawker board minutes and accounts, Sopwith board minutes (no accession numbers).

Rolls-Royce, Derby
Rolls-Royce archive: Ernest Hives papers, Sir Arthur Sidgreaves papers, Rolls-Royce board minutes, record of Rolls-Royce engines despatched to 31 December 1961 (no accession numbers).

Engineering Employers Federation, London
Engineering Employers Federation archive (including case records): Fiche references A (1) 30, 32, 38, 43, 46, 50, 52, 55, A (6) 41; D (1) 45, 145, 139, D (8) 42; M (14) 13, M (19) 69; P (5) 188, 222, 225; S (9) 12.

Royal Aeronautical Society Library, London:
J.V. Connolly papers (no accession numbers).

Official publications

United States Strategic Bombing Survey, Report 4, *Aircraft Division Industry Report (European Theater)*.
United States Strategic Bombing Survey, Report 54, *Machine Tools and Machinery as Capital Equipment*.
Public Accounts Committee, Second Report (London, 1944).
Annual Statements of Trade of UK.
Census of Production.

Interviews and correspondence

Letter from A.E. Tagg (formerly of Hawker) to the author, 12 December 1990.

Author's interviews with S.D. Davies (Hawker design staff, 1931–36, assistant chief designer, A.V. Roe, 1938–44), 21 and 30 April 1990; letter from S.D. Davies to the author, 8 May 1990; 'Personal history of S.D. Davies at A.V. Roe (Manchester) 1938–44', autobiographical account of S.D. Davies' work at A.V. Roe, sent to the author in June 1990.

Books

C. Andrews and E. Morgan, *Supermarine Aircraft Since 1914* (London, 1987).

W. Ashworth, *Contracts and Finance* (London, 1953).

F. Banks, *I Kept No Diary* (Shrewsbury, 1978).

C. Barnes, *Bristol Aircraft Since 1910* (London 1964).

C. Barnett, *The Audit of War: The Illusion and Reality of Britain as a Great Nation* (London, 1986).

C. Barnett, *The Collapse of British Power* (London, 1972).

P.M.H. Bell, *The Origins of the Second World War in Europe* (London, 1986).

N.K. Buxton and D.H. Aldcroft (eds), *British Industry Between the Wars: Instability and Industrial Development, 1919–39* (London, 1979).

Sir Alec Cairncross, *Planning in Wartime: Aircraft Production in Britain, Germany and America* (London, 1991).

D. Chester (ed.), *Lessons of the British War Economy* (Cambridge, 1951).

R. Church, *Herbert Austin: The British Motor Industry to 1941* (London, 1979).

B. Collins and R. Robbins (eds), *British Culture and Economic Decline* (London, 1991).

J. Cross, *Lord Swinton* (Oxford, 1982).

R. Croucher, *Engineers at War, 1939–1945* (London, 1982).

J. Davy, *The Standard Car, 1903–63* (Coventry, 1965).

E. Devons, *Planning in Practice: Essays in Aircraft Planning in Wartime* (Cambridge, 1950).

Dictionary of National Biography.

M. Donne, *Pioneers of the Skies: A History of Short Brothers Plc* (Belfast, 1987).

D. Edgerton, *England and the Aeroplane: An Essay on a Militant and Technological Nation* (London, 1991).

B. Elbaum and W. Lazonick (eds), *The Decline of the British Economy* (Oxford, 1986).

C. Gibbs-Smith, *Aviation: An Historical Survey from its Origins to the End of World War II* (London, 1970).

M. Gilbert, *Winston S. Churchill, V, 1922–1939* (London, 1976).

M. Gilbert, *Winston S. Churchill, VI, 1939–1941* (London, 1983).

G. Gordon, *British Seapower and Procurement between the Wars: A Reappraisal of Rearmament* (London, 1988).

B. Gunston, *By Jupiter! The Life of Sir Roy Fedden* (London, 1978).

H. Hall and C. Wrigley, *Studies in Overseas Supply* (London, 1956).

R. Hall, *North American Supply* (London, 1955).

W. Hancock and M. Gowing, *The British War Economy* (London, 1949).

L. Hannah, *The Rise of the Corporate Economy* (London, 1976).

A. Harvey-Bailey, *Rolls-Royce: Hives, the Quiet Tiger* (Sir Henry Royce Memorial Foundation, 1985).

A. Harvey-Bailey, *The Merlin in Perspective: The Combat Years* (Derby, Rolls-Royce Heritage Trust, 1983).

I. Holley, jr., *Buying Aircraft: Materiel Procurement for the Army Air Forces* (Washington, 1964).

Sir Stanley Hooker (assisted by B. Gunston), *Not Much of an Engineer: an Autobiography* (Shrewsbury, 1984).

W. Hornby, *Factories and Plant* (London, 1958).

J. Hurstfield, *The Control of Raw Materials* (London, 1953).

P. Inman, *Labour in the Munitions Industry* (London, 1957).

A. Jackson, *Avro Aircraft Since 1908* (London, 1965).

A. Jackson, *Blackburn Aircraft Since 1909* (London, 1968).

D. James, *Gloster Aircraft Since 1917* (London, 1971).

P. King, *Knights of the Air* (London, 1989).

K. Hayward, *The British Aircraft Industry* (Manchester, 1989).

H. Montgomery Hyde, *British Air Policy between the Wars* (London, 1976).

I. Lloyd, *Rolls-Royce: The Merlin at War* (London, 1978).

I. Lloyd, *Rolls-Royce: The Years of Endeavour* (London, 1978).

R. Locke, *The End of the Practical Man: Entrepreneurship and Higher Education in Germany, France and Great Britain, 1880–1940* (Greenwich CT, 1984).

F. Mason, *Hawker Aircraft Since 1920* (London, 1971).

W. Manser, *Britain in Balance: the Myth of Failure* (Harmondsworth, 1971).

Sir Eric Mensforth, *Family Engineers* (London, 1981).

A. Milward, *War, Economy and Society, 1939–45* (London, 1977).

R. Miller and D. Sawers, *The Technical Development of Modern Aviation* (London, 1968).

J. Morpurgo, *Barnes Wallis* (London, 1972).

R. Overy, *The Air War* (New York, 1981).

R. Overy, *Goering: The Iron Man* (London, 1984).

R. Overy with A. Wheatcroft, *The Road to War* (London, 1989).

R. Overy, *William Morris, Viscount Nuffield* (London, 1976).

H. Parker, *Manpower: A Study of Wartime Policy and Administration* (London, 1957).

G. Peden, *British Rearmament and the Treasury 1932–1939* (Edinburgh, 1979).

S. Pollard, *The Development of the British Economy 1914–1992* (London, 1992).

S. Pollard, *The Wasting of the British Economy* (London, 1982).

M. Postan, *British War Production* (London, 1952).

M. Postan, D. Hay and J. Scott, *The Design and Development of Weapons* (London, 1964).

W. Reader, *Architect of Air Power: The Life of the first Viscount Weir of Eastwood, 1877–1959* (London, 1968).

M. Sanderson, *The Universities and British Industry* (London, 1972).

R. Sayers, *Financial Policy, 1939–45* (London, 1956).

R. Schlaifer and R. Heron, *The Development of Aircraft Engines and Aviation Fuels* (Boston MA, 1950).

J. Scott, *Vickers: A History* (London, 1962).

J. Scott and R. Hughes, *The Administration of War Production* (London, 1955).

C. Martin Sharp, *D.H. – A History of De Havilland* (London, 1960).

R. Shay, *British Rearmament in the Thirties* (Princeton, 1977).

N. Shute, *Slide Rule: The Autobiography of an Engineer* (London, 1954).

A. Slaven and D. Aldcroft (eds), *Business, Banking and Urban History: Essays in Honour of S.G. Checkland* (Edinburgh, 1982).

M. Smith, *British Air Strategy between the Wars* (Oxford, 1984).

Statistical Digest of the War (London, 1951).

Lord Swinton, *I Remember* (London, 1948).

A. Taylor, *Beaverbrook* (London, 1972).

H. Taylor, *Airspeed Aircraft since 1931* (London, 1970).

H. Taylor, *Fairey Aircraft Since 1915* (London, 1988).

J. Terraine, *The Right of the Line* (London, 1988).

D. Thoms, *War, Industry and Society: The Midlands, 1939–45* (London, 1989).

D. Thoms and T. Donnelly, *The Motor Industry in Coventry since the 1890s* (London, 1985).

H. Tuffen and A. Tagg, *The Hawker Hurricane: Design, Development and Production* (Royal Aeronautical Society Historical Group, 1985).

M. Wiener, *English Culture and the Decline of the Industrial Spirit, 1850–1980* (Cambridge, 1981).

C. Wilson and W. Reader, *Men and Machines: A History of D. Napier and Son Ltd., 1808–1958* (London, 1958).

Articles

Anon., 'Building the Packard-Merlin', *The Archive*, 3 (1989).

Sir Alec Cairncross, 'Reflections: How British Aircraft Production was Planned in the Second World War', *Twentieth Century British History*, Vol. 2, No. 3 (1991).

F. Coghlan, 'Armaments, Economic Policy and Appeasement: Background to British Foreign Policy, 1931–1937', *History*, 57 (1972).

S. Davies, 'Aeroplane Design and Production', *Aircraft Engineering*, 11 (1939).

R. Dobson and R. Taylor, 'The Jointing of Materials by Welding', *Journal of the Royal Aeronautical Society*, 40 (1936).

D. Edgerton, 'The Prophet Militant and Industrial: The Peculiarities of Correlli Barnett', *Twentieth Century British History*, Vol. 2, No. 3 (1991).

D. Edgerton, 'Science and Technology in British Business History', *Business History* 29 (1987).

D. Edgerton, 'Technical Innovation, Industrial Capacity and Efficiency: Public Ownership and the British Military Aircraft Industry, 1935–48', *Business History*, 26 (1984).

C. Fairey, 'The Future of Aeroplane Design for the Services', *Royal United Services Institution Journal*, 76 (1931).

P. Fearon, 'The Formative Years of the British Aircraft Industry, 1913–1924', *Business History Review*, 43 (1969).

P. Fearon, 'The British Airframe Industry and the State, 1918–35', *Economic History Review*, 27 (1974).

P. Fearon, 'The Vicissitudes of a British Aircraft Company: Handley Page Ltd Between the Wars', *Business History*, 20 (1978).

J. Ferris, 'Treasury Control: the Ten Year Rule and British Service Policies, 1919–1924', *Historical Journal*, 30 (1987).

M. Gowing, 'The Organisation of Manpower in Britain during the Second World War', *Journal of Contemporary History*, 7 (1972).

R. Higham, 'Government, Companies and National Defense: British Aeronautical Experience 1918–1945 as the basis of a broad hypothesis', *Business History Review*, 39 (1965).

R. Higham, 'Quantity vs. Quality: The Impact of Changing Demand on the British Aircraft Industry, 1900–1960', *Business History Review*, 42 (1968).

S. Marriner, 'Company Financial Statements as Source Material for Business Historians', *Business History*, 22 (1980).

E. Mensforth, 'Airframe Production', *Proceedings of the Institution of Mechanical Engineers*, 156 (1947).

E. Mensforth and W. Petter, 'Aspects of the Design and Production of Airframes with Particular Reference to their Co-ordination and to the Reduction of the Development Period', *Journal of the Royal Aeronautical Society*, 48 (1944).

F. Miles, 'Track Assembly', *Flight*, 38 (August 1940).

G. Mills, 'Ford and the Merlin', *The Archive*, 9 (1991).

R. Overy, 'The German Pre-war Aircraft Production Plans: November 1936–April 1939', *English Historical Review*, 90 (1975).

R. Parker, 'British Rearmament 1936–1939: Treasury, trade unions and skilled labour', *English Historical Review*, 96 (1981).

G. Peden, 'A Matter of Timing: The Economic Background to British Foreign Policy, 1938–1939', *History*, 69 (1984).

A. Robertson, 'The British Airframe Industry and the State in the Interwar Period: A Comment', *Economic History Review*, 28 (1975).

A. Robertson, 'British Rearmament and Industrial Growth, 1935–39', *Research in Economic History*, 8 (1983).

D. Robinson, 'Some Developments in Aircraft Production', *Journal of the Royal Aeronautical Society*, 53 (1949).

M. Smith, 'Planning and Building the British Bomber Force, 1934–1939', *Business History Review*, 54 (1980).

M. Thomas, 'Rearmament and Economic Recovery in the Late 1930s', *Economic History Review*, 36 (1983).

T. Wright, 'Factors Affecting the Cost of Airplanes', *Journal of Aeronautical Sciences*, 3 (1936).

Periodicals

The Statist
Air Force List
Aircraft Production
The Aeroplane

Index of aircraft and aero-engines

General Index